PRAISE FOR *SOMETHING DEEPLY HIDDEN*

"A thrilling tour through what is perhaps humankind's greatest intellectual achievement—quantum mechanics. With bold clarity, Carroll deftly unmasks quantum weirdness to reveal a strange but utterly wondrous reality."

—Brian Greene, professor of physics and mathematics, director of the Columbia University Center for Theoretical Physics, and author of *The Elegant Universe*

"What makes Carroll's new project so worthwhile, though, is that while he is most certainly choosing sides in the debate, he offers us a cogent, clear, and compelling guide to the subject while letting his passion for the scientific questions shine through every page." —NPR

"Carroll gives us a front-row seat to the development of a new vision of physics: one that connects our everyday experiences to a dizzying hall-of-mirrors universe in which our very sense of self is challenged. It's a fascinating idea, and one that just might hold clues to a deeper reality."

—Katie Mack, theoretical astrophysicist at North Carolina State University and author of *The End of Everything*

"Enlightening and refreshingly bold." —*Scientific American*

"Sean Carroll's immensely enjoyable *Something Deeply Hidden* brings readers face-to-face with the fundamental quantum weirdness of the universe—or should I say universes? And by the end, you may catch yourself finding quantum weirdness not all that weird."

—Jordan Ellenberg, professor of mathematics at University of Wisconsin–Madison and author of *How Not to Be Wrong*

"*Something Deeply Hidden* is Carroll's ambitious and engaging foray into what quantum mechanics really means and what it tells us about physical reality." ...e

"Sean Carroll is always lucid and funny, gratifyingly readable, while still excavating depths. He advocates an acceptance of quantum mechanics at its most minimal, its most austere—appealing to the allure of the pristine. The consequence is an annihilation of our conventional notions of reality in favor of an utterly surreal world of Many Worlds. Sean includes us in the battle between a simple reality versus a multitude of realities that feels barely on the periphery of human comprehension. He includes us in the ideas, the philosophy, and the foment of revolution. A fascinating and important book."

—Janna Levin, professor of physics and astronomy at Barnard College of Columbia University and author of *Black Hole Blues and Other Songs from Outer Space*

"Carroll argues with a healthy restlessness that makes his book more interesting than so many others in the quantum physics genre."

—*Forbes*

"Sean Carroll beautifully clarifies the debate about the foundations of quantum mechanics and champions the most elegant, courageous approach: the astonishing 'Many Worlds' interpretation. His explanations of its pros and cons are clear, evenhanded, and philosophically gobsmacking."

—Steven Strogatz, professor of mathematics at Cornell University and author of *Infinite Powers*

"If you want to know why some people take [the Everett] approach seriously and what you can do with it, then Carroll's latest is one of the best popular books on the market."

—*Physics Today*

"I was overwhelmed by tears of joy at seeing so many fundamental issues explained as well as they ever have been. *Something Deeply Hidden* is a masterpiece, which stands along with Feynman's *QED* as one of the two best popularizations of quantum mechanics I've ever seen. And if we classify *QED* as having had different goals, then it's just the best popularization of quantum mechanics I've ever seen, full stop."

—Scott Aaronson, professor of computer science at the University of Texas at Austin and director of UT Austin's Quantum Information Center

"Be prepared to deal with some equations—and to have your mind blown."

—GeekWire

"Irresistible and an absolute treat to read. While this is a book about some of the deepest current mysteries in physics, it is also a book about metaphysics, as Carroll lucidly guides us on how to not only think about the true and hidden nature of reality but also how to make sense of it. I loved this book."

—Priyamvada Natarajan, theoretical astrophysicist at Yale University and author of *Mapping the Heavens*

"By far the most articulate and cogent defense of the Many-Worlds view in book-length depth with a close connection to the latest ongoing research."

—*Science News*

"Solid arguments and engaging historical backdrop will captivate science-minded readers everywhere."

—*Scientific Inquirer*

"As a smart and intensely readable undergraduate class in the history of quantum theory and the nature of quantum mechanics, *Something Deeply Hidden* could scarcely be improved."

—Steve Donoghue, *Open Letters Monthly*

"Readers in this universe (and others?) will relish the opportunity to explore the frontiers of science in the company of titans."

—*Booklist*

"Fans of popular science authors such as Neil deGrasse Tyson and John Gribbin will find great joy while exploring these groundbreaking concepts."

—*Library Journal*

"[A] challenging, provocative book . . . moving smoothly through different topics and from objects as small as particles to those as enormous as black holes, Carroll's exploration of quantum theory introduces readers to some of the most groundbreaking ideas in physics today."

—*Publishers Weekly*

ALSO BY SEAN CARROLL

From Eternity to Here

The Particle at the End of the Universe

The Big Picture

SOMETHING DEEPLY HIDDEN

Quantum Worlds and the Emergence of Spacetime

SEAN CARROLL

DUTTON

DUTTON

An imprint of Penguin Random House LLC
penguinrandomhouse.com

Previously published as a Dutton hardcover in 2019
First trade paperback printing: September 2020

LIBRARY OF CONGRESS CATALOGING-IN-PUBLICATION DATA
has been applied for.

Dutton trade paperback ISBN: 9781524743031

Printed in the United States of America
5th Printing

BOOK DESIGN BY LAURA K. CORLESS

Interior art: distressed overlay texture © gooddesign10 / Shutterstock.com
Version number 756,132,390,815,553

To thinkers throughout history

who stuck to their guns

for the right reasons

CONTENTS

SOMETHING DEEPLY HIDDEN

PROLOGUE

Don't Be Afraid

You don't need a PhD in theoretical physics to be afraid of quantum mechanics. But it doesn't hurt.

That might seem strange. Quantum mechanics is our best theory of the microscopic world. It describes how atoms and particles interact through the forces of nature, and makes incredibly precise experimental predictions. To be sure, quantum mechanics has something of a reputation for being difficult, mysterious, just this side of magic. But professional physicists, of all people, should be relatively comfortable with a theory like that. They are constantly doing intricate calculations involving quantum phenomena, and building giant machines dedicated to testing the resulting predictions. Surely we're not suggesting that physicists have been faking it all this time?

They haven't been faking, but they haven't exactly been honest with themselves either. On the one hand, quantum mechanics is the heart and soul of modern physics. Astrophysicists, particle physicists, atomic physicists, laser physicists—everyone uses quantum mechanics all the time, and they're very good at it. It's not just a matter of esoteric research. Quantum mechanics is ubiquitous in modern technology.

Semiconductors, transistors, microchips, lasers, and computer memory all rely on quantum mechanics to function. For that matter, quantum mechanics is necessary to make sense of the most basic features of the world around us. Essentially all of chemistry is a matter of applied quantum mechanics. To understand how the sun shines, or why tables are solid, you need quantum mechanics.

Imagine closing your eyes. Hopefully things look pretty dark. You might think that makes sense, because no light is coming in. But that's not quite right; infrared light, with a slightly longer wavelength than visible light, is being emitted all the time by any warm object, and that includes your own body. If our eyes were as sensitive to infrared light as they are to visible light, we would be blinded even when our lids were closed, from all the light emitted by our eyeballs themselves. But the rods and cones that act as light receptors in our eyes are cleverly sensitive to visible light, not infrared. How do they manage that? Ultimately, the answer comes down to quantum mechanics.

Quantum mechanics isn't magic. It is the deepest, most comprehensive view of reality we have. As far as we currently know, quantum mechanics isn't just an approximation of the truth; it is the truth. That's subject to change in the face of unexpected experimental results, but we've seen no hints of any such surprises thus far. The development of quantum mechanics in the early years of the twentieth century, involving names like Planck, Einstein, Bohr, Heisenberg, Schrödinger, and Dirac, left us by 1927 with a mature understanding that is surely one of the greatest intellectual accomplishments in human history. We have every reason to be proud.

On the other hand, in the memorable words of Richard Feynman, "I think I can safely say that nobody understands quantum mechanics." We *use* quantum mechanics to design new technologies and predict the outcomes of experiments. But honest physicists admit that we don't truly *understand* quantum mechanics. We have a recipe that we can safely apply in certain prescribed situations, and which returns mind-bogglingly precise predictions that have been triumphantly vindicated by the data. But if you want to dig deeper and ask what is really going on, we simply don't know. Physicists tend to treat quantum mechanics like a mindless robot they rely on to perform certain tasks, not as a beloved friend they care about on a personal level.

This attitude among the professionals seeps into how quantum mechanics gets explained to the wider world. What we would like to do is to present a fully formed picture of Nature, but we can't quite do that, since physicists don't agree about what quantum mechanics actually says. Instead, popular treatments tend to emphasize that quantum mechanics is mysterious, baffling, impossible to understand. That message goes against the basic principles that science stands for, which include the idea that the world is fundamentally intelligible. We have something of a mental block when it comes to quantum mechanics, and we need a bit of quantum therapy to help get past it.

○ ○ ○

When we teach quantum mechanics to students, they are taught a list of rules. Some of the rules are of a familiar type: there's a mathematical description of quantum systems, plus an explanation of how such systems evolve over time. But then there are a bunch of extra rules that have no analogue in any other theory of physics. These extra rules tell us what happens when we *observe* a quantum system, and that behavior is completely different from how the system behaves when we're not observing it. What in the world is going on with that?

There are basically two options. One is that the story we've been telling our students is woefully incomplete, and in order for quantum mechanics to qualify as a sensible theory we need to understand what a "measurement" or "observation" is, and why it seems so different from what the system does otherwise. The other option is that quantum mechanics represents a violent break from the way we have always thought about physics before, shifting from a view where the world exists objectively and independently of how we perceive it, to one where the act of observation is somehow fundamental to the nature of reality.

In either case, the textbooks should by all rights spend time exploring these options, and admit that even though quantum mechanics is extremely successful, we can't claim to be finished developing it just yet. They don't. For the most part, they pass over this issue in silence, preferring to stay in the physicist's comfort zone of writing down equations and challenging students to solve them.

That's embarrassing. And it gets worse.

You might think, given this situation, that the quest to understand quantum mechanics would be the single biggest goal in all of physics. Millions of dollars of grant money would flow to researchers in quantum foundations, the brightest minds would flock to the problem, and the most important insights would be rewarded with prizes and prestige. Universities would compete to hire the leading figures in the area, dangling superstar salaries to lure them away from rival institutions.

Sadly, no. Not only is the quest to make sense of quantum mechanics not considered a high-status specialty within modern physics; in many quarters it's considered barely respectable at all, if not actively disparaged. Most physics departments have nobody working on the problem, and those who choose to do so are looked upon with suspicion. (Recently while writing a grant proposal, I was advised to concentrate on describing my work in gravitation and cosmology, which is considered legitimate, and remain silent about my work on the foundations of quantum mechanics, as that would make me appear less serious.) There have been important steps forward over the last ninety years, but they have typically been made by headstrong individuals who thought the problems were important despite what all of their colleagues told them, or by young students who didn't know any better and later left the field entirely.

In one of Aesop's fables, a fox sees a juicy bunch of grapes and leaps to reach it, but can't quite jump high enough. In frustration he declares that the grapes were probably sour, and he never really wanted them anyway. The fox represents "physicists," and the grapes are "understanding quantum mechanics." Many researchers have decided that understanding how nature really works was never really important; all that matters is the ability to make particular predictions.

Scientists are trained to value tangible results, whether they are exciting experimental findings or quantitative theoretical models. The idea of working to understand a theory we already have, even if that effort might not lead to any specific new technologies or predictions, can be a tough sell. The underlying tension was illustrated in the TV show *The Wire*, where a group of hardworking detectives labored for months to meticulously gather evidence that would build a case against a powerful drug ring. Their bosses, meanwhile, had no patience for such incremental frivolity. They just wanted to see drugs on the table for their next press conference, and encouraged the police to bang heads and make splashy arrests. Funding agencies and hiring committees are like those bosses. In a world where all the incentives push us toward concrete, quantifiable outcomes, less pressing big-picture concerns can be pushed aside as we race toward the next immediate goal.

° ° °

This book has three main messages. The first is that quantum mechanics should be understandable—even if we're not there yet—and achieving such understanding should be a high-priority goal of modern science. Quantum mechanics is unique among physical theories in drawing an apparent distinction between *what we see* and *what really is*. That poses a special challenge to the minds of scientists (and everyone else), who are used to thinking about what we see as unproblematically "real," and working to explain things accordingly. But this challenge isn't insuperable, and if we free our minds from certain old-fashioned and intuitive ways of thinking, we find that quantum mechanics isn't hopelessly mystical or inexplicable. It's just physics.

The second message is that we have made real progress toward understanding. I will focus on the approach I feel is clearly the most promising route, the Everett or Many-Worlds formulation of quantum mechanics. Many-Worlds has been enthusiastically embraced by many physicists, but it has a sketchy reputation among people who are put off by a proliferation of other realities containing copies of themselves. If you are one of those people, I want to at least convince you that Many-Worlds is the *purest* way of making sense of quantum mechanics—it's where we end up if we just follow the path of least resistance in taking quantum phenomena seriously. In particular, the multiple worlds are predictions of the formalism that is already in place, not something added in by hand. But Many-Worlds isn't the only respectable approach, and we will mention some of its main competitors. (I will endeavor to be fair, if not necessarily balanced.) The important thing is that the various approaches are all well-constructed scientific theories, with potentially different experimental ramifications, not just woolly-headed "interpretations" to be debated over cognac and cigars after we're finished doing real work.

The third message is that all this matters, and not just for the integrity of science. The success to date of the existing adequate-but-not-perfectly-coherent framework of quantum mechanics shouldn't blind us to the fact that there are circumstances under which such an approach simply isn't up to the task. In particular, when we turn to understanding the nature of spacetime itself, and the origin and ultimate fate of the entire universe, the foundations of quantum mechanics are absolutely crucial. I'll introduce some new, exciting, and admittedly tentative proposals that draw provocative connections between quantum entanglement and how spacetime bends and curves—the phenomenon you and I know as "gravity." For many years now, the search for a complete and compelling quantum theory of gravity has been recognized as an important scientific goal (prestige, prizes, stealing away faculty, and all that). It may be that the secret is not to start with gravity and "quantize" it, but to dig deeply into quantum mechanics itself, and find that gravity was lurking there all along.

We don't know for sure. That's the excitement and anxiety of cutting-edge research. But the time has come to take the fundamental nature of reality seriously, and that means confronting quantum mechanics head-on.

Part One

SPOOKY

1

What's Going On

Looking at the Quantum World

Albert Einstein, who had a way with words as well as with equations, was the one who stuck quantum mechanics with the label it has been unable to shake ever since: *spukhaft*, usually translated from German to English as "spooky." If nothing else, that's the impression we get from most public discussions of quantum mechanics. We're told that it's a part of physics that is unavoidably mystifying, weird, bizarre, unknowable, strange, baffling. Spooky.

Inscrutability can be alluring. Like a mysterious, sexy stranger, quantum mechanics tempts us into projecting all sorts of qualities and capacities onto it, whether they are there or not. A brief search for books with "quantum" in the title reveals the following list of purported applications:

Quantum Success
Quantum Leadership
Quantum Consciousness
Quantum Touch

Quantum Yoga
Quantum Eating
Quantum Psychology
Quantum Mind
Quantum Glory
Quantum Forgiveness
Quantum Theology
Quantum Happiness
Quantum Poetry
Quantum Teaching
Quantum Faith
Quantum Love

For a branch of physics that is often described as only being relevant to microscopic processes involving subatomic particles, that's a pretty impressive résumé.

To be fair, quantum mechanics—or "quantum physics," or "quantum theory," the labels are all interchangeable—is not only relevant to microscopic processes. It describes the whole world, from you and me to stars and galaxies, from the centers of black holes to the beginning of the universe. But it is only when we look at the world in extreme close-up that the apparent weirdness of quantum phenomena becomes unavoidable.

One of the themes in this book is that quantum mechanics doesn't deserve the connotation of spookiness, in the sense of some ineffable mystery that it is beyond the human mind to comprehend. Quantum mechanics is *amazing*; it is novel, profound, mind-stretching, and a very different view of reality from what we're used to. Science is like that sometimes. But if the subject seems difficult or puzzling, the scientific response is to solve the puzzle, not to pretend it's not there. There's every reason to think we can do that for quantum mechanics just like any other physical theory.

Many presentations of quantum mechanics follow a typical pattern. First, they point to some counterintuitive quantum phenomenon. Next, they express bafflement that the world can possibly be that way, and despair of it making sense. Finally (if you're lucky), they attempt some sort of explanation.

Our theme is prizing clarity over mystery, so I don't want to adopt that strategy. I want to present quantum mechanics in a way that will make it maximally understandable right from the start. It will still seem strange, but that's the nature of the beast. What it won't seem, hopefully, is inexplicable or unintelligible.

We will make no effort to follow historical order. In this chapter we'll look at the basic experimental facts that force quantum mechanics upon us, and in the next we'll quickly sketch the Many-Worlds approach to making sense of those observations. Only in the chapter after that will we offer a semi-historical account of the discoveries that led people to contemplate such a dramatically new kind of physics in the first place. Then we'll hammer home exactly how dramatic some of the implications of quantum mechanics really are.

With all that in place, over the rest of the book we can set about the fun task of seeing where all this leads, demystifying the most striking features of quantum reality.

○ ○ ○

Physics is one of the most basic sciences, indeed one of the most basic human endeavors. We look around the world, we see it is full of stuff. What is that stuff, and how does it behave?

These are questions that have been asked ever since people started asking questions. In ancient Greece, physics was thought of as the general study of change and motion, of both living and nonliving matter. Aristotle spoke a vocabulary of tendencies, purposes, and causes. How an entity moves and changes can be explained by reference to its inner nature and to external powers acting upon it. Typical objects, for example, might by nature be at rest; in order for them to move, it is necessary that something be causing that motion.

All of this changed thanks to a clever chap named Isaac Newton. In 1687 he published *Principia Mathematica*, the most important work in the history of physics. It was there that he laid the groundwork for what we now call "classical" or simply "Newtonian" mechanics. Newton blew away any dusty talk of natures and purposes, revealing what lay underneath: a crisp, rigorous mathematical formalism with which teachers continue to torment students to this very day.

Whatever memory you may have of high-school or college homework assignments dealing with pendulums and inclined planes, the basic ideas of classical mechanics are pretty simple. Consider an object such as a rock. Ignore everything about the rock that a geologist might consider interesting, such as its color and composition. Put aside the possibility that the basic structure of the rock might change, for example, if you smashed it to pieces with a hammer. Reduce your mental image of the rock down to its most abstract form: the rock is an object, and that object has a *location in space*, and that location *changes with time*.

Classical mechanics tells us precisely how the position of the rock changes with time. We're very used to that by now, so it's worth reflecting on how impressive this is. Newton doesn't hand us some vague platitudes about the general tendency of rocks to move more or less in this or that fashion. He gives us exact, unbreakable rules for how everything in the universe moves in response to everything else—rules that can be used to catch baseballs or land rovers on Mars.

Here's how it works. At any one moment, the rock will have a position and also a velocity, a rate at which it's moving. According to Newton, if no forces act on the rock, it will continue to move in a straight line at constant velocity, for all time. (Already this is a major departure from Aristotle, who would have told you that objects need to be constantly pushed if they are to be kept in motion.) If a force does act on the rock, it will cause acceleration—some change in the velocity of the rock, which might make it go faster, or slower, or merely alter its direction—in direct proportion to how much force is applied.

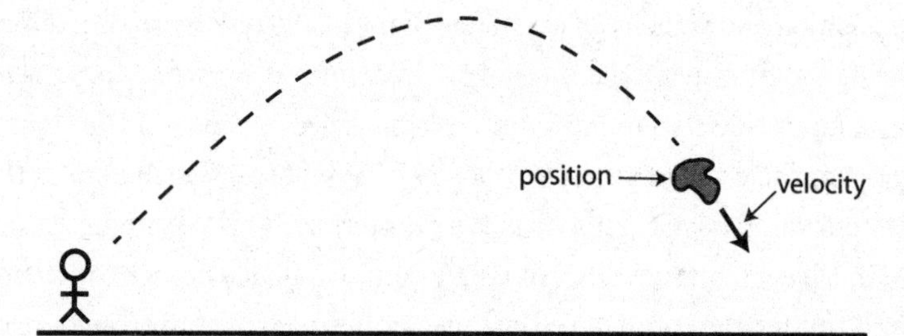

That's basically it. To figure out the entire trajectory of the rock, you need to tell me its position, its velocity, and what forces are acting on it. Newton's equations tell you the rest. Forces might include the force of gravity, or the force of your hand if you pick up the rock and throw it, or the force from the ground when the rock comes to land. The idea works just as well for billiard balls or rocket ships or planets. The project of physics, within this classical paradigm, consists essentially of figuring out what makes up the stuff of the universe (rocks and so forth) and what forces act on them.

Classical physics provides a straightforward picture of the world, but a number of crucial moves were made along the way to setting it up. Notice that we had to be very specific about what information we required to figure out what would happen to the rock: its position, its velocity, and the forces acting on it. We can think of those forces as being part of the outside world, and the important information about the rock itself as consisting of just its position and velocity. The acceleration of the rock at any moment in time, by contrast, is not something we need to specify; that's exactly what Newton's laws allow us to calculate from the position and the velocity.

Together, the position and velocity make up the *state* of any object in classical mechanics. If we have a system with multiple moving parts, the classical state of that entire system is just a list of the states of each of the individual parts. The air in a normal-sized room will have perhaps 10^{27} molecules of different types, and the state of that air would be a list of the position and velocity of every one of them. (Strictly speaking, physicists like to use the momentum of each particle, rather than its velocity, but as far as Newtonian mechanics is concerned the momentum is simply the particle's mass times its velocity.) The set of all possible states that a system could have is known as the *phase space* of the system.

The French mathematician Pierre-Simon Laplace pointed out a profound implication of the classical mechanics way of thinking. In principle, a vast intellect could know the state of literally every object in the universe, from which it could deduce everything that would happen in the future, as well as everything that had happened in the past. *Laplace's demon* is a thought experiment, not a realistic project for an ambitious computer scientist, but the implications of the thought experiment are profound. Newtonian mechanics describes a deterministic, clockwork universe.

The machinery of classical physics is so beautiful and compelling that it seems almost inescapable once you grasp it. Many great minds who came after Newton were convinced that the basic superstructure of physics had been solved, and future progress lay in figuring out exactly what realization of classical physics (which particles, which forces) was the right one to describe the universe as a whole. Even relativity, which was world-transforming in its own way, is a variety of classical mechanics rather than a replacement for it.

Then along came quantum mechanics, and everything changed.

∘ ∘ ∘

Alongside Newton's formulation of classical mechanics, the invention of quantum mechanics represents the other great revolution in the history of physics. Unlike anything that had come before, quantum theory didn't propose a particular physical model within the basic classical framework; it discarded that framework entirely, replacing it with something profoundly different.

The fundamental new element of quantum mechanics, the thing that makes it unequivocally distinct from its classical predecessor, centers on the question of what it means to *measure* something about a quantum system. What exactly a measurement is, and what happens when we measure something, and what this all tells us about what's really happening behind the scenes: together, these questions constitute what's called the *measurement problem* of quantum mechanics. There is absolutely no consensus within physics or philosophy on how to solve the measurement problem, although there are a number of promising ideas.

Attempts to address the measurement problem have led to the emergence of a field known as *the interpretation of quantum mechanics*, although the label isn't very accurate. "Interpretations" are things that we might apply to a work of literature or art, where people might have different ways of thinking about the same basic object. What's going on in quantum mechanics is something else: a competition between truly distinct scientific theories, incompatible ways of making sense of the physical world. For this reason, modern workers in this field prefer to call it "foundations of quantum mechanics." The subject of quantum foundations is part of science, not literary criticism.

Nobody ever felt the need to talk about "interpretations of classical mechanics"—classical mechanics is perfectly transparent. There is a mathematical formalism that speaks of positions and velocities and trajectories, and oh, look: there is a rock whose actual motion in the world obeys the predictions of that formalism. There is, in particular, no such thing as a measurement problem in classical mechanics. The state of the system is given by its position and its velocity, and if we want to measure those quantities, we simply do so. Of course, we can measure the system sloppily or crudely, thereby obtaining imprecise results or altering the system itself. But we don't have to; just by being careful, we can precisely measure everything there is to know about the system without altering it in any noticeable way. Classical mechanics offers a clear and unambiguous relationship between what we see and what the theory describes.

Quantum mechanics, for all its successes, offers no such thing. The enigma at the heart of quantum reality can be summed up in a simple motto: what we *see* when we look at the world seems to be fundamentally different from what actually *is*.

◦ ◦ ◦

Think about electrons, the elementary particles orbiting atomic nuclei, whose interactions are responsible for all of chemistry and hence almost everything interesting around you right now. As we did with the rock, we can ignore some of the electron's specific properties, like its spin and the fact that it has an electric field. (Really we could just stick with the rock as our example—rocks are quantum systems just as much as electrons are—but switching to a subatomic particle helps us remember that the features distinguishing quantum mechanics only become evident when we consider very tiny objects indeed.)

Unlike in classical mechanics, where the state of a system is described by its position and velocity, the nature of a quantum system is something a bit less concrete. Consider an electron in its natural habitat, orbiting the nucleus of an atom. You might think, from the word "orbit" as well as from the numerous cartoon depictions of atoms you have doubtless been exposed to over the years, that the orbit of an electron is more or less like the orbit of a planet in the solar system. The electron (so you might think) has a location, and a velocity, and as time passes it zips around the central nucleus in a circle or maybe an ellipse.

Quantum mechanics suggests something different. We can *measure* values of the location or velocity (though not at the same time), and if we are sufficiently careful and talented experimenters we will obtain some answer. But what we're seeing through such a measurement is not the actual, complete, unvarnished state of the electron. Indeed, the particular measurement outcome we will obtain cannot be predicted with perfect confidence, in a profound departure from the ideas of classical mechanics. The best we can do is to predict the *probability* of seeing the electron in any particular location or with any particular velocity.

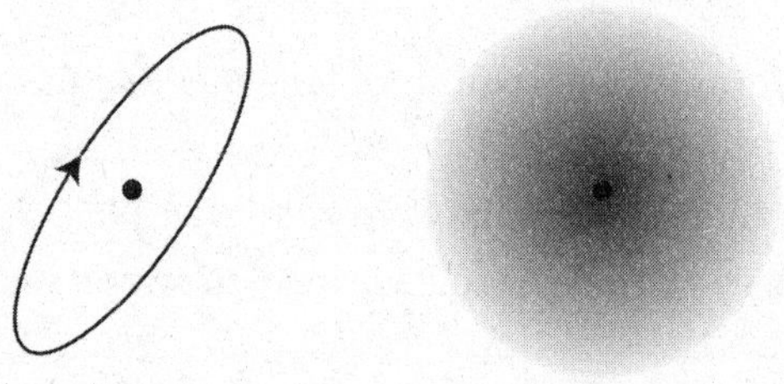

Classical electron orbit

Quantum electron wave function

The classical notion of the state of a particle, "its location and its velocity," is therefore replaced in quantum mechanics by something utterly alien to our everyday experience: a cloud of probability. For an electron in an atom, this cloud is more dense toward the center and thins out as we get farther away. Where the cloud is thickest, the probability of seeing the electron is highest; where it is diluted almost to imperceptibility, the probability of seeing the electron is vanishingly small.

This cloud is often called a *wave function*, because it can oscillate like a wave, as the most probable measurement outcome changes over time. We usually denote a wave function by Ψ, the Greek letter Psi. For every possible measurement outcome, such as the position of the particle, the wave function assigns a specific number, called the *amplitude* associated with that outcome. The amplitude that a particle is at some position x_0, for example, would be written $\Psi(x_0)$.

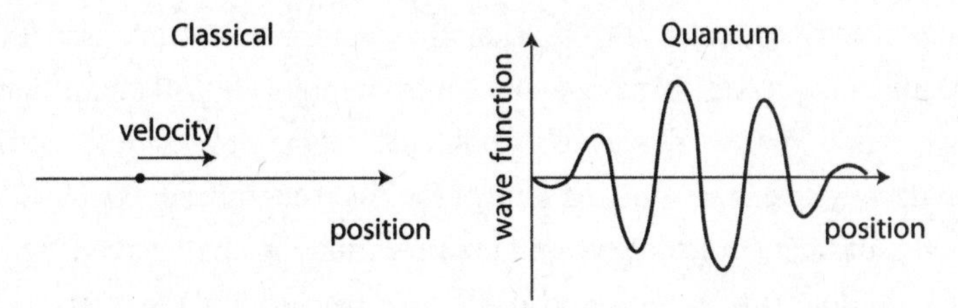

The probability of getting that outcome when we perform a measurement is given by the amplitude squared.

Probability of a particular outcome = |Amplitude for that outcome|^2

This simple relation is called the *Born rule*, after physicist Max Born.* Part of our task will be to figure out where in the world such a rule came from.

We're most definitely *not* saying that there is an electron with some position and velocity, and we just don't know what those are, so the wave function encapsulates our ignorance about those quantities. In this chapter we're not saying anything at all about what "is," only what we observe. In chapters to come, I will pound the table and insist that the wave function is the sum total of reality, and ideas such as the

* There's a slight technicality, which we'll mention here and then pretty much forget about: the amplitude for any given outcome is actually a complex number, not a real number. Real numbers are the ones that appear on the number line, any number between minus infinity and plus infinity. Anytime you take the square of a real number, you get another real number that is greater than or equal to zero, so as far as real numbers are concerned there's no such thing as the square root of a negative number. Mathematicians long ago realized that square roots of negative numbers would be really useful things to have, so they defined the "imaginary unit" i as the square root of -1. An imaginary number is just a real number, called "the imaginary part," times i. Then a complex number is just a combination of a real number and an imaginary one. The little bars in the notation |Amplitude|^2 in the Born rule mean that we actually add the squares of the real and the imaginary parts. All that is just for the sticklers out there; henceforth we'll be happy to say "the probability is the amplitude squared" and be done with it.

position or the velocity of the electron are merely things we can measure. But not everyone sees things that way, and for the moment we are choosing to don a mask of impartiality.

° ° °

Let's place the rules of classical and quantum mechanics side by side to compare them. The state of a classical system is given by the position and velocity of each of its moving parts. To follow its evolution, we imagine something like the following procedure:

Rules of Classical Mechanics

1. Set up the system by fixing a specific position and velocity for each part.
2. Evolve the system using Newton's laws of motion.

That's it. The devil is in the details, of course. Some classical systems can have a lot of moving pieces.

In contrast, the rules of standard textbook quantum mechanics come in two parts. In the first part, we have a structure that exactly parallels that of the classical case. Quantum systems are described by wave functions rather than by positions and velocities. Just as Newton's laws of motion govern the evolution of the state of a system in classical mechanics, there is an equation that governs how wave functions evolve, called *Schrödinger's equation*. We can express Schrödinger's equation in words as: "The rate of change of a wave function is proportional to the energy of the quantum system." Slightly more specifically, a wave function can represent a number of different possible energies, and the Schrödinger equation says that high-energy parts of the wave function evolve rapidly, while low-energy parts evolve very slowly. Which makes sense, when we think about it.

What matters for our purposes is simply that there is such an equation, one that predicts how wave functions evolve smoothly through time. That evolution is as predictable and inevitable as the way objects move according to Newton's laws in classical mechanics. Nothing weird is happening yet.

The beginning of the quantum recipe reads something like this:

Rules of Quantum Mechanics (Part One)

1. Set up the system by fixing a specific wave function Ψ.
2. Evolve the system using Schrödinger's equation.

So far, so good—these parts of quantum mechanics exactly parallel their classical predecessors. But whereas the rules of classical mechanics stop there, the rules of quantum mechanics keep going.

All the extra rules deal with measurement. When you perform a measurement, such as the position or spin of a particle, quantum mechanics says there are only certain possible results you will ever get. You can't predict which of the results it will be, but you can calculate the probability for each allowed outcome. And after your measurement is done, the wave function *collapses* to a completely different function, with all of the new probability concentrated on whatever result you just got. So if you measure a quantum system, in general the best you can do is predict probabilities for various outcomes, but if you were to immediately measure the same quantity again, you will always get the same answer—the wave function has collapsed onto that outcome.

Let's write this out in gory detail.

Rules of Quantum Mechanics (Part Two)

3. There are certain observable quantities we can choose to measure, such as position, and when we do measure them, we obtain definite results.

4. The probability of getting any one particular result can be calculated from the wave function. The wave function associates an amplitude with every possible measurement outcome; the probability for any outcome is the square of that amplitude.
5. Upon measurement, the wave function collapses. However spread out it may have been pre-measurement, afterward it is concentrated on the result we obtained.

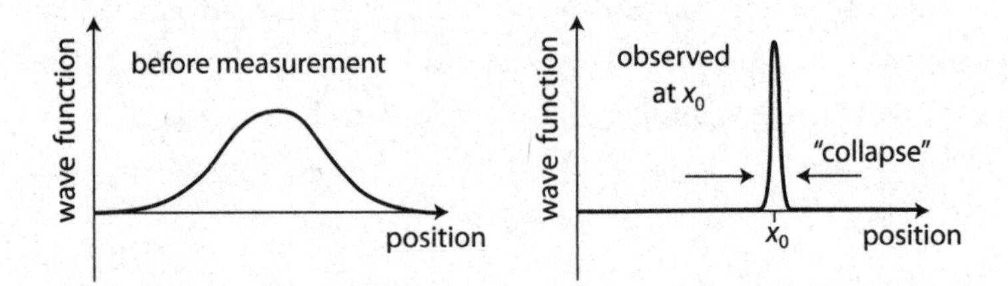

In a modern university curriculum, when physics students are first exposed to quantum mechanics, they are taught some version of these five rules. The ideology associated with this presentation—treat measurements as fundamental, wave functions collapse when they are observed, don't ask questions about what's going on behind the scenes—is sometimes called the *Copenhagen interpretation* of quantum mechanics. But people, including the physicists from Copenhagen who purportedly invented this interpretation, disagree on precisely what that label should be taken to describe. We can just refer to it as "standard textbook quantum mechanics."

The idea that these rules represent how reality actually works is, needless to say, outrageous.

What precisely do you mean by a "measurement"? How quickly does it happen? What exactly constitutes a measuring apparatus? Does it need to be human, or have some amount of consciousness, or perhaps the ability to encode information? Or maybe it just has to be macroscopic, and if so how macroscopic does it have to be? When exactly does the measurement occur, and how quickly? How in the world does the wave function collapse so dramatically? If the wave function were very spread out, does the collapse happen faster than the speed of light? And what happens to all the possibilities that were seemingly allowed by the wave function but which we didn't observe? Were they never really there? Do they just vanish into nothingness?

To put things most pointedly: Why do quantum systems evolve smoothly and deterministically according to the Schrödinger equation *as long as we aren't looking at them*, but then dramatically collapse when we do look? How do they know, and why do they care? (Don't worry, we're going to answer all these questions.)

◦ ◦ ◦

Science, most people think, seeks to understand the natural world. We observe things happening, and science hopes to provide an explanation for what is going on.

In its current textbook formulation, quantum mechanics has failed in this ambition. We don't know what's really going on, or at least the community of professional physicists cannot agree on what it is. What we have instead is a *recipe* that we enshrine in textbooks and teach to our students. Isaac Newton could tell you, starting with the position and velocity of a rock that you have thrown into the air in the Earth's gravitational field, just what the subsequent trajectory of that rock was going to be. Analogously, starting with a quantum system prepared in some particular way, the rules of quantum mechanics can tell you how the wave function will change over time, and what the probability of various possible measurement outcomes will be should you choose to observe it.

The fact that the quantum recipe provides us with probabilities rather that certainties might be annoying, but we could learn to live with it. What bugs us, or should, is our lack of understanding about what is actually happening.

Imagine that some devious genius figured out all the laws of physics, but rather than revealing them to the rest of the world, they programmed a computer to answer questions concerning specific physics problems, and put an interface to the program on a web page. Anyone who was interested could just surf over to that site, type in a well-posed physics question, and get the correct answer.

Such a program would obviously be of great use to scientists and engineers. But having access to the site wouldn't qualify as understanding the laws of physics. We would have an oracle that was in the business of providing answers to specific questions, but we ourselves would be completely lacking in any intuitive idea of the underlying rules of the game. The rest of the world's scientists, presented with such an oracle, wouldn't be moved to declare victory; they would continue with their work of figuring out what the laws of nature actually were.

Quantum mechanics, in the form in which it is currently presented in physics textbooks, represents an oracle, not a true understanding. We can set up specific problems and answer them, but we can't honestly explain what's happening behind the scenes. What we do have are a number of good ideas about what that could be, and it's past time that the physics community started taking these ideas seriously.

The Courageous Formulation

Austere Quantum Mechanics

The attitude inculcated into young students by modern quantum mechanics textbooks has been compactly summarized by physicist N. David Mermin as "Shut up and calculate!" Mermin himself wasn't advocating such a position, but others have. Every decent physicist spends a good deal of time calculating things, whatever their attitude toward quantum foundations might be. So really the admonition could be shortened to simply "Shut up!"*

It wasn't always thus. Quantum mechanics took decades to piece together, but rounded into its modern form around 1927. In that year, at the Fifth International Solvay Conference in Belgium, the world's

* If you look on the Internet, you will find numerous attributions of "Shut up and calculate!" to Richard Feynman, a physicist who was an all-time great at doing difficult calculations. But he never said any such thing, nor would he have found the sentiment congenial; Feynman thought carefully about quantum mechanics, and nobody ever accused him of shutting up. It's common for quotations to be reattributed to plausible speakers who are more famous than the actual source of the quote. Sociologist Robert Merton has dubbed this the Matthew Effect, after a line from the Gospel of Matthew: "For unto every one that hath shall be given, and he shall have abundance: but from him that hath not shall be taken away even that which he hath."

leading physicists came together to discuss the status and meaning of quantum theory. By that time the experimental evidence was clear, and physicists were at long last in possession of a quantitative formulation of the rules of quantum mechanics. It was time to roll up some sleeves and figure out what this crazy new worldview actually amounted to.

The discussions at this conference help set the stage, but our goal here isn't to get the history right. We want to understand the physics. So we'll sketch out a logical path by which we will be led to a full-blown scientific theory of quantum mechanics. No vague mysticism, no seemingly ad hoc rules. Just a simple set of assumptions leading to some remarkable conclusions. With this picture in mind, many things that might otherwise have seemed ominously mysterious will suddenly start to make perfect sense.

○ ○ ○

The Solvay Conference has gone down in history as the beginning of a famous series of debates between Albert Einstein and Niels Bohr over how to think about quantum mechanics. Bohr, a Danish physicist based in Copenhagen who is rightfully regarded as the godfather of quantum theory, advocated an approach similar to the textbook recipe we discussed in the last chapter: use quantum mechanics to calculate the probabilities for measurement outcomes, but don't ask of it anything more than that. Do not, in particular, worry too much about what is really happening behind the scenes. Supported by his younger colleagues Werner Heisenberg and Wolfgang Pauli, Bohr insisted that quantum mechanics was a perfectly fine theory as it was.

Einstein would have none of it. He was firmly convinced that the duty of physics was precisely to ask what was going on behind the scenes, and that the state of quantum mechanics in 1927 fell far short of providing a satisfactory account of nature. With his own sympathizers, such as Erwin Schrödinger and Louis de Broglie, Einstein advocated looking more deeply, and attempting to extend and generalize quantum mechanics into a satisfactory physical theory.

Participants in the 1927 Solvay Conference. Among the more well-known were: 1. Max Planck, 2. Marie Curie, 3. Paul Dirac, 4. Erwin Schrödinger, 5. Albert Einstein, 6. Louis de Broglie, 7. Wolfgang Pauli, 8. Max Born, 9. Werner Heisenberg, and 10. Niels Bohr. (Courtesy of Wikipedia)

Einstein and his compatriots had reason to be cautiously optimistic that such a new-and-improved theory was out there to be found. Just a few decades before, in the later years of the nineteenth century, physicists had developed the theory of statistical mechanics, which described the motion of large numbers of atoms and molecules. A key step in that development—which all took place under the rubric of classical mechanics, before quantum theory came on the scene—was the idea that we can talk profitably about the behavior of a large collection of particles even if we don't know precisely the position and velocity of each one of them. All we need to know is a *probability distribution* describing the likelihood that the particles might be behaving in various ways.

In statistical mechanics, in other words, we think that there actually is some particular classical state of all the particles, but we don't know it, all we have is a distribution of probabilities. Happily, such a distribution is all we need to do a great deal of useful physics, since it fixes properties such as the temperature and pressure of the system. But the distribution isn't a complete description of the system; it's simply a reflection of what we know (or don't) about it. To tag this distinction with philosophical buzzwords, in statistical mechanics the probability distribution is an *epistemic* notion—describing the state of our knowledge—rather than an *ontological* one—describing some objective feature of reality. Epistemology is the study of knowledge; ontology is the study of what is real.

It was natural, in 1927, to suspect that quantum mechanics should be thought of along similar lines. After all, by that time we had figured out that what we use wave functions for is to calculate the probability of any particular measurement outcome. Surely it makes sense to imagine that nature itself knows precisely what the outcome is going to be, but the formalism of quantum theory simply doesn't completely capture that knowledge, and thus needs to be improved. The wave function, in this view, isn't the whole story; there are additional "hidden variables" that fix what the actual measurement outcomes are going to be, even if we don't know (and perhaps can't ever determine ahead of the measurement) what their values are.

Maybe. But in subsequent years a number of results have been obtained, most notably by the physicist John Bell in the 1960s, implying that the most simple and straightforward attempts along these lines are doomed to failure. People tried—de Broglie actually put forward a specific theory, which was rediscovered and extended by David Bohm in the 1950s, and Einstein and Schrödinger both batted around ideas. But Bell's theorem implies that any such theory requires "action at a distance"—a measurement at one location can instantly affect the state of the universe arbitrarily far away. This seems to be in violation of the spirit if not the letter of the theory of relativity, which says that objects and influences cannot propagate faster than the speed of light. The hidden-variable approach is still being actively pursued, but all known attempts along these lines are ungainly and hard to reconcile with modern theories such as the Standard Model of particle physics, not to mention speculative ideas about quantum gravity, as we'll discuss later. Perhaps this is why Einstein, the pioneer of relativity, never found a satisfactory theory of his own.

In the popular imagination, Einstein lost the Bohr-Einstein debates. We are told that Einstein, a creative revolutionary in his youth, had grown old and conservative, and was unable to accept or even understand the dramatic implications of the new quantum theory. (At the time of the Solvay Conference Einstein was forty-eight years old.) Physics subsequently went on without him, as the great man retreated to pursue idiosyncratic attempts at finding a unified field theory.

Nothing could be further from the truth. While Einstein failed to put forward a complete and compelling generalization of quantum mechanics, his insistence that physics needs to do better than shut up and calculate was directly on point. It is wildly off base to think that he failed to understand quantum theory. Einstein understood it as well as anyone, and continued to make fundamental contributions to the subject, including demonstrating the importance of quantum entanglement, which plays a central role in our current best picture of how the universe really works. What he failed to do was to convince his fellow physicists of the inadequacy of the Copenhagen approach, and the importance of trying harder to understand the foundations of quantum theory.

◦ ◦ ◦

If we want to follow Einstein's ambition of a complete, unambiguous, realistic theory of the natural world, but we are discouraged by the difficulties of tacking new hidden variables onto quantum mechanics, is there any remaining strategy left?

One approach is to forget about new variables, throw away all the problematic ideas about the measurement process, strip quantum mechanics down to its absolute essentials, and ask what happens. What's the leanest, meanest version of quantum theory we can invent, and still hope to explain the experimental results?

Every version of quantum mechanics (and there are plenty) employs a wave function or something equivalent, and posits that the wave function obeys Schrödinger's equation, at least most of the time. These are going to have to be ingredients in just about any theory we can take seriously. Let's see if we can be stubbornly minimalist, and get away with adding little or nothing else to the formalism.

This minimalist approach has two aspects. First, we take the wave function seriously as a direct representation of reality, not just a bookkeeping device to help us organize our knowledge. We treat it as ontological, not epistemic. That's the most austere strategy we can imagine adopting, since anything else would posit additional structure over and above the wave function. But it's also a dramatic step, since wave functions are very different from what we observe when we look at the world. We don't *see* wave functions; we see measurement outcomes, like the position of a particle. But the theory seems to demand that wave functions play a central role, so let's see how far we can get by imagining that reality is exactly described by a quantum wave function.

Second, if the wave function usually evolves smoothly in accordance with the Schrödinger equation, let's suppose that's what it always does. In other words, let's erase all of those extra rules about measurement in the quantum recipe entirely, and bring things back to the stark simplicity of the classical paradigm: there is a wave function, and it evolves according to a deterministic rule, and that's all there is to say. We might call this proposal "austere quantum mechanics," or AQM for short. It stands in contrast with textbook quantum mechanics, where we appeal to collapsing wave functions and try to avoid talking about the fundamental nature of reality altogether.

A bold strategy. But there's an immediate problem with it: it sure *seems* like wave functions collapse. When we make measurements of a quantum system with a spread-out wave function, we get a specific answer. Even if we think an electron wave function is a diffuse cloud centered on the nucleus, when we actually look at it we don't see such a cloud, we see a point-like particle at some particular location. And if we look immediately again, we see the electron in basically the same location. There's a good reason why the pioneers of quantum mechanics invented the idea of wave functions collapsing—because that's what they appear to do.

But maybe that's too quick. Let's turn the question around. Rather than starting with what we see and trying to invent a theory to explain it, let's start with austere quantum mechanics (wave functions evolving smoothly, that's it), and ask what people in a world described by that theory would actually experience.

Think about what this could mean. In the last chapter, we were careful to talk about the wave function as a kind of mathematical black box from which predictions for measurement outcomes could be extracted: for any particular outcome, the wave function assigns an amplitude, and the probability of getting that outcome is the amplitude squared. Max Born, who proposed the Born rule, was one of the attendees at Solvay in 1927.

Now we're saying something deeper and more direct. The wave function isn't a bookkeeping device; it's an exact representation of the quantum system, just as a set of positions and velocities would be a representation of a classical system. The world *is* a wave function, nothing more nor less. We can use the phrase "quantum state" as a synonym for "wave function," in direct parallel with calling a set of positions and velocities a "classical state."

This is a dramatic claim about the nature of reality. In ordinary conversation, even among grizzled veterans of quantum physics, people are always talking about concepts like "the position of the electron." But this wave-function-is-everything view implies that such talk is wrongheaded in an important way. There is no such thing as "the position of the electron." There is only the electron's wave function. Quantum mechanics implies a profound distinction between "what we can observe" and "what there really is." Our observations aren't revealing preexisting facts of which we were previously ignorant; at best, they reveal a tiny slice of a much bigger, fundamentally elusive reality.

Consider an idea you will often hear: "Atoms are mostly empty space." Utterly wrong, according to the AQM way of thinking. It comes from a stubborn insistence on thinking of an electron as a tiny classical dot zipping around inside of the wave function, rather than the electron actually *being* the wave function. In AQM, there's nothing zipping around; there is only the quantum state. Atoms aren't mostly empty space; they are described by wave functions that stretch throughout the extent of the atom.

The way to break out of our classical intuition is to truly abandon the idea that the electron has some particular location. An electron is in a *superposition* of every possible position we could see it in, and it doesn't snap into any one specific location until we actually observe it to be there. "Superposition" is the word physicists use to emphasize that the electron exists in a combination of all positions, with a particular amplitude for each one. Quantum reality is a wave function; classical positions and velocities are merely what we are able to observe when we probe that wave function.

◦ ◦ ◦

So the reality of a quantum system, according to austere quantum mechanics, is described by a wave function or quantum state, which can be thought of as a superposition of every possible outcome of some observation we might want to make. How do we get from there to the annoying reality that wave functions appear to collapse when we make such measurements?

Start by examining the statement "we measure the position of the electron" a little more carefully. What does this measurement process actually involve? Presumably some lab equipment and a bit of experimental dexterity, but we don't need to worry about specifics. All we need to know is that there is some measuring apparatus (a camera or whatever) that somehow interacts with the electron, and then lets us read off where the electron was seen.

In the textbook quantum recipe, that's as much insight as we would ever get. Some of the people who pioneered this approach, including Niels Bohr and Werner Heisenberg, would go a little bit further, making explicit the idea that the measuring apparatus should be thought of as a classical object, even if the electron it was observing was quantum-mechanical. This line of division between the parts of the world that should be treated using quantum versus classical descriptions is sometimes called the *Heisenberg cut.* Rather than accepting that quantum mechanics is fundamental and classical mechanics is just a good approximation to it in appropriate circumstances, textbook quantum mechanics puts the classical world at center stage, as the right way to talk

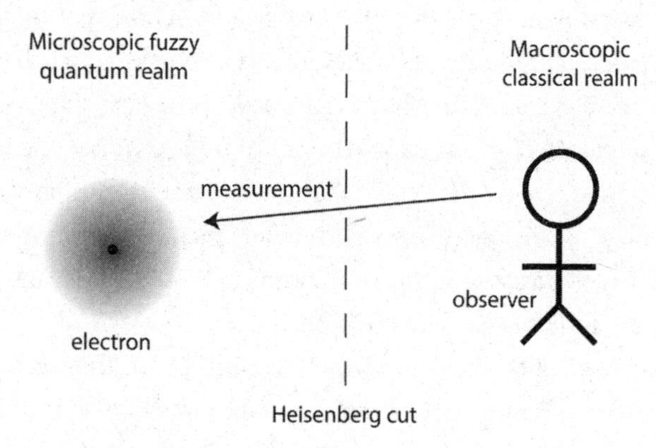

about people and cameras and other macroscopic things that interact with microscopic quantum systems.

This doesn't smell right. One's first guess should be that the quantum/classical divide is a matter of our personal convenience, not a fundamental aspect of nature. If atoms obey the rules of quantum mechanics and cameras are made of atoms, presumably cameras obey the rules of quantum mechanics too. For that matter, you and I presumably obey the rules of quantum mechanics. The fact that we are big, lumbering, macroscopic objects might make classical physics a good approximation to what we are, but our first guess should be that it's really quantum from top to bottom.

If that's true, it's not just the electron that has a wave function. The camera should have a wave function of its own. So should the experimenter. Everything is quantum.

That simple shift of perspective suggests a new angle on the measurement problem. The AQM attitude is that we shouldn't treat the measurement process as anything mystical or even in need of its own set of rules; the camera and the electron simply interact with each other according to the laws of physics, just like a rock and the earth do.

A quantum state describes systems as superpositions of different measurement outcomes. The electron will, in general, start out in a superposition of various locations—all the places we could see it were we to look. The camera starts out in some wave function that might look complicated, but amounts to saying "This is a camera, and it hasn't yet looked at the electron." But then it does look at the electron, which is a physical interaction governed by the Schrödinger equation. And after that interaction, we might expect that the camera itself is now in a superposition of all the possible measurement outcomes it might have observed: the camera saw the electron in *this* location, or the camera saw the electron in *that* location, and so on.

If that were the whole story, AQM would be an untenable mess. Electrons in superpositions, cameras in superpositions, nothing much resembling the robust approximately classical world of our experience.

Fortunately we can appeal to another startling feature of quantum mechanics: given two different objects (like an electron and a camera), they are not described by separate, individual wave functions. There is *only one wave function*, which describes the entire system we care about, all the way up to the "wave function of the universe" if we're talking about the whole shebang. In the case under consideration, there is a wave function describing the combined electron+camera system. So what we really have is a superposition of all possible combinations of where the electron might have been located, and where the camera actually observed it to be.

Although such a superposition in principle includes every possibility, most of the possible outcomes are assigned zero weight in the quantum state. The cloud of probability vanishes into nothingness for most possible combinations of electron location and camera image. In particular, there is no probability that the electron was in one location but the camera saw it somewhere else (as long as you have a relatively functional camera).

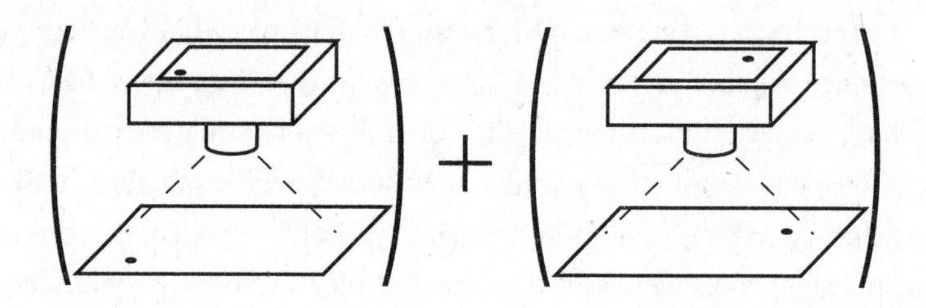

This is the quantum phenomenon known as *entanglement*. There is a single wave function for the combined electron+camera system, consisting of a superposition of various possibilities of the form "the electron was at this location, and the camera observed it at the same location." Rather than the electron and the camera doing their own thing, there is a connection between the two systems.

Now let's take every appearance of "camera" in the above discussion and replace it with "you." Rather than taking a picture with a mechanical apparatus, we (fancifully) imagine that you have really good eyesight and can see where electrons are just by looking at them. Otherwise, nothing changes. According to the Schrödinger equation, an initially unentangled situation—the electron is in a superposition of various possible locations, and you haven't looked at the electron yet—evolves smoothly into an entangled one—a superposition of each location the electron could have been observed, and you having seen the electron in just that location.

That's what the rules of quantum mechanics would say, if we hadn't tacked on all of those extra annoying bits about the measurement process. Maybe all of those extra rules were just a waste of time. In AQM, the story we just told, about you and the electron entangling and evolving into a superposition, is the complete story. There isn't anything special about measurement; it's just something that happens when two systems interact in an appropriate way. And afterward, *you and the system you interacted with are in a superposition*, in each part of which you have seen the electron in a slightly different location.

The problem is, this story still doesn't match onto what you actually experience when you observe a quantum system. You never feel like you have evolved into a superposition of different possible measurement outcomes; you simply think you've seen some specific outcome, which can be predicted with a definite probability. That's why all of those extra measurement rules were added in the first place. Otherwise you seemingly have a very pretty and elegant formalism (quantum states, smooth evolution) that just doesn't match up to reality.

○ ○ ○

Time to get a little philosophical. What exactly do we mean by "you" in the above paragraph? Constructing a scientific theory isn't simply a matter of writing down some equations; we also need to indicate how those equations map onto the world. When it comes to you and me, we tend to think that the process of matching ourselves onto some part of a scientific formalism is pretty straightforward. Certainly in the story told above, where an observer measures the position of an electron, it definitely seems as if that observer evolves into an entangled superposition of the different possible measurement outcomes.

But there's an alternative possibility. Before the measurement happened, there was one electron and one observer (or camera, if you prefer—it doesn't matter how we think about the thing that interacts with the electron as long as it's a big, macroscopic object). After they interact, however, rather than thinking of that one observer having evolved into a superposition of possible states, we could think of them as having evolved into *multiple possible observers*. The right way to describe things after the measurement, in this view, is not as one person with multiple ideas about where the electron was seen, but as *multiple worlds*, each of which contains a single person with a very definite idea about where the electron was seen.

Here's the big reveal: what we've described as austere quantum mechanics is more commonly known as the Everett, or Many-Worlds, formulation of quantum mechanics, first put forward by Hugh Everett in 1957. The Everett view arises from a fundamental annoyance with all of the special rules about measurements that are presented as part of the standard textbook quantum recipe, and suggests instead that there is just a single kind of quantum evolution. The price we pay for this vastly increased elegance of theoretical formalism is that the theory describes many copies of what we think of as "the universe," each slightly different, but each truly real in some sense. Whether the benefit is worth the cost is an issue about which people disagree. (It is.)

In stumbling upon the Many-Worlds formulation, at no point did we take ordinary quantum mechanics and tack on a bunch of universes. The potential for such universes was always there—the universe has a wave function, which can very naturally describe superpositions of many different ways things could be, including superpositions of the whole universe. All we did is to point out that this potential is naturally actualized in the course of ordinary quantum evolution. Once you admit that an electron can be in a superposition of different locations, it follows that a person can be in a superposition of having seen the electron in different locations, and indeed that reality as a whole can be in a superposition, and it becomes natural to treat every term in that superposition as a separate "world." We didn't add anything to quantum mechanics, we just faced up to what was there all along.

We might reasonably call Everett's approach the "courageous" formulation of quantum mechanics. It embodies the philosophy that we should take seriously the simplest version of underlying reality that accounts for what we see, even if that reality differs wildly from our everyday experience. Do we have the courage to accept it?

◦ ◦ ◦

This brief introduction to Many-Worlds leaves many questions unanswered. When exactly does the wave function split into many worlds? What separates the worlds from one another? How many worlds are there? Are the other worlds really "real"? How would we ever know, if we can't observe them? (Or can we?) How does this explain the probability that we'll end up in one world rather than another one?

All of these questions have good answers—or at least plausible ones—and much of the book to come will be devoted to answering them. But we should also admit that the whole picture might be wrong, and something very different is required.

Every version of quantum mechanics features two things: (1) a wave function, and (2) the Schrödinger equation, which governs how wave functions evolve in time. The entirety of the Everett formulation is simply the insistence that there is *nothing else*, that these ingredients suffice to provide a complete, empirically adequate account of the world. ("Empirically adequate" is a fancy way that philosophers like to say "it fits the data.") Any other approach to quantum mechanics consists of adding something to that bare-bones formalism, or somehow modifying what is there.

The most immediately startling implication of pure Everettian quantum mechanics is the existence of many worlds, so it makes sense to call it Many-Worlds. But the essence of the theory is that reality is described by a smoothly evolving wave function and nothing else. There are extra challenges associated with this philosophy, especially when it comes to matching the extraordinary simplicity of the formalism to the rich diversity of the world we observe. But there are corresponding advantages of clarity and insight. As we'll see when we ultimately turn to quantum field theory and quantum gravity, taking wave functions as primary in their own right, free of any baggage inherited from our classical experience, is extraordinarily helpful when tackling the deep problems of modern physics.

Given the necessity of these two ingredients (wave functions and the Schrödinger equation), there are a few alternatives to Many-Worlds we might also consider. One is to imagine adding new physical entities over and above the wave function. This approach leads to hidden-variable models, which were in the back of the minds of people like Einstein from the start. These days the most popular such approach is called the *de Broglie–Bohm theory*, or simply *Bohmian mechanics*. Alternatively, we could leave the wave function by itself but imagine changing the Schrödinger equation, for example, to introduce real, random collapses. Finally, we might imagine that the wave function isn't a physical thing at all, but simply a way of characterizing what we know about reality. Such approaches are broadly known as epistemic models, and a currently popular version is *QBism*, or *quantum Bayesianism*.

All of these options—and there are many more not listed here—represent truly distinct physical theories, not simply "interpretations" of the same underlying idea. The existence of multiple incompatible theories that all lead (at least thus far) to the observable predictions of quantum mechanics creates a conundrum for anyone who wants to talk about what quantum theory really means. While the quantum recipe is agreed upon by working scientists and philosophers, the underlying reality—what any particular phenomenon actually *means*—is not.

I am defending one particular view of that reality, the Many-Worlds version of quantum mechanics, and for most of this book I will simply be explaining things in Many-Worlds terms. This shouldn't be taken to imply that the Everettian view is unquestionably right. I hope to explain what the theory says, and why it's reasonable to assign a high credence to it being the best view of reality we have; what you personally end up believing is up to you.

Why Would Anybody Think This?

How Quantum Mechanics Came to Be

"Sometimes I've believed as many as six impossible things before breakfast," notes the White Queen to Alice in *Through the Looking Glass*. That can seem like a useful skill as one comes to grips with quantum mechanics in general, and Many-Worlds in particular. Fortunately, the impossible-seeming things we're asked to believe aren't whimsical inventions or logic-busting Zen koans; they are features of the world that we are nudged toward accepting because actual experiments have dragged us, kicking and screaming, in that direction. We don't choose quantum mechanics; we only choose to face up to it.

Physics aspires to figure out what kinds of stuff the world is made of, how that stuff naturally changes over time, and how various bits of stuff interact with one another. In my own environment, I can immediately see many different kinds of stuff: papers and books and a desk and a computer and a cup of coffee and a wastebasket and two cats (one of whom is extremely interested in what's inside the wastebasket), not to mention less solid things like air and light and sound.

By the end of the nineteenth century, scientists had managed to distill every single one of these things down to two fundamental kinds of substances: *particles* and *fields*. Particles are point-like objects at a definite location in space, while fields (like the gravitational field) are spread throughout space, taking on a particular value at every point. When a field is oscillating across space and time, we call that a "wave." So people will often contrast particles with waves, but what they really mean is particles and fields.

Quantum mechanics ultimately unified particles and fields into a single entity, the wave function. The impetus to do so came from two directions: first, physicists discovered that things they thought were waves, like the electric and magnetic fields, had particle-like properties. Then they realized that things they thought were particles, like electrons, manifested field-like properties. The reconciliation of these puzzles is that the world is fundamentally field-like (it's a quantum wave function), but when we look at it by performing a careful measurement, it looks particle-like. It took a while to get there.

○ ○ ○

Particles seem to be pretty straightforward things: objects located at particular points in space. The idea goes back to ancient Greece, where a small group of philosophers proposed that matter was made up of point-like "atoms," for the Greek word for "indivisible." In the words of Democritus, the original atomist, "Sweet is by convention, bitter by convention, hot by convention, cold by convention, color by convention; in truth there are only atoms and the void."

At the time there wasn't that much actual evidence in favor of the proposal, so it was largely abandoned until the beginning of the 1800s, when experimenters had begun to study chemical reactions in a quantitative way. A crucial role was played by tin oxide, a compound made of tin and oxygen, which was discovered to come in two different forms.

The English scientist John Dalton noted that for a fixed amount of tin, the amount of oxygen in one form of tin oxide was exactly twice the amount in the other. We could explain this, Dalton argued in 1803, if both elements came in the form of discrete particles, for which he borrowed the word "atom" from the Greeks. All we have to do is to imagine that one form of tin oxide was made of single tin atoms combined with single oxygen atoms, while the other form consisted of single tin atoms combined with two oxygen atoms. Every kind of chemical element, Dalton suggested, was associated with a unique kind of atom, and the tendency of the atoms to combine in different ways was responsible for all of chemistry. A simple summary, but one with world-altering implications.

Dalton jumped the gun a little bit with his nomenclature. For the Greeks, the whole point of atoms was that they were indivisible, the fundamental building blocks out of which everything else is made. But Dalton's atoms are not at all indivisible—they consist of a compact nucleus surrounded by orbiting electrons. It took over a hundred years to realize that, however. First the English physicist J. J. Thomson discovered electrons in 1897. These seemed to be an utterly new kind of particle, electrically charged and only 1/1800th the mass of hydrogen, the lightest atom. In 1909 Thomson's former student Ernest Rutherford, a New Zealand physicist who had moved to the UK for his advanced studies, showed most of the mass of the atom was concentrated in a central nucleus, while the atom's overall size was set by the orbits of much lighter electrons traveling around that nucleus. The standard cartoon picture of an atom, with electrons circling the nucleus much like planets orbit the sun in our solar system, represents this Rutherford model of atomic structure. (Rutherford didn't know about quantum mechanics, so this cartoon deviates from reality in significant ways, as we shall see.)

Further work, initiated by Rutherford and followed up by others,

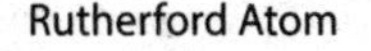

Rutherford Atom

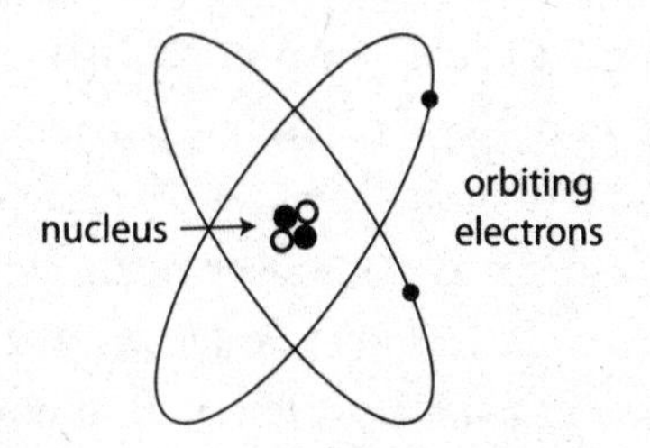

revealed that nuclei themselves aren't elementary, but consist of positively charged protons and uncharged neutrons. The electric charges of electrons and protons are equal in magnitude but opposite in sign, so an atom with an equal number of each (and however many neutrons you like) will be electrically neutral. It wasn't until the 1960s and '70s that physicists established that protons and neutrons are also made of smaller particles, called quarks, held together by new force-carrying particles called gluons.

Chemically speaking, electrons are where it's at. Nuclei give atoms their heft, but outside of rare radioactive decays or fission/fusion reactions, they basically go along for the ride. The orbiting electrons, on the other hand, are light and jumpy, and their tendency to move around is what makes our lives interesting. Two or more atoms can share electrons, leading to chemical bonds. Under the right conditions, electrons can change their minds about which atoms they want to be associated with, which gives us chemical reactions. Electrons can even escape their atomic captivity altogether in order to move freely through a substance, a phenomenon we call "electricity." And when you shake an electron, it sets up a vibration in the electric and magnetic fields around it, leading to light and other forms of electromagnetic radiation.

To emphasize the idea of being truly point-like, rather than a small object but with some definite nonzero size, we sometimes distinguish between "elementary" particles, which define literal points in space, and "composite" particles that are really made of even smaller constituents. As far as anyone can tell, electrons are truly elementary particles. You can see why discussions of quantum mechanics are constantly referring to electrons when they reach for examples—they're the easiest fundamental particle to make and manipulate, and play a central role in the behavior of the matter of which we and our surroundings are made.

° ° °

In bad news for Democritus and his friends, nineteenth-century physics didn't explain the world in terms of particles alone. It suggested, instead, that two fundamental kinds of stuff were required: both particles and fields.

Fields can be thought of as the opposite of particles, at least in the context of classical mechanics. The defining feature of a particle is that it's located at one point in space, and nowhere else. The defining feature of a field is that it is located everywhere. A field is something that has a value at literally every point in space. Particles need to interact with each other somehow, and they do so through the influence of fields.

Think of the magnetic field. It's a *vector field*—at every point in space it looks like a little arrow, with a magnitude (the field can be strong, or weak, or even exactly zero) and also a direction (it points along some particular axis). We can measure the direction in which the magnetic field points just by pulling out a magnetic compass and observing what direction the needle points in. (It will point roughly north, if you are located at most places on Earth and not standing too close to another magnet.) The important thing is that the magnetic field exists invisibly everywhere throughout space, even when we're not observing it. That's what fields do.

There is also the electric field, which is also a vector with a magnitude and a direction at every point in space. Just as we can detect a magnetic field with a compass, we can detect the electric field by placing an electron at rest and seeing if it accelerates. The faster the acceleration, the stronger the electric field.* One of the triumphs of nineteenth-century physics was when James Clerk Maxwell unified electricity and magnetism, showing that both of these fields could be thought of as different manifestations of a single underlying "electromagnetic" field.

The other field that was well known in the nineteenth century is the gravitational field. Gravity, Isaac Newton taught us, stretches over astronomical distances. Planets in the solar system feel a gravitational pull toward the sun, proportional to the sun's mass and inversely proportional to the square of the distance between them. In 1783 Pierre-Simon Laplace showed that we can think of Newtonian gravity as arising from a "gravitational potential field" that has a value at every point in space, just as the electric and magnetic fields do.

○ ○ ○

By the end of the 1800s, physicists could see the outlines of a complete theory of the world coming into focus. Matter was made of atoms, which were made of smaller particles, interacting via various forces carried by fields, all operating under the umbrella of classical mechanics.

What the World Is Made Of (Nineteenth-Century Edition)

- Particles (point-like, making up matter).
- Fields (pervading space, giving rise to forces).

---

* Annoyingly, the electron accelerates in precisely the opposite direction that the electric field points, because by human convention we've decided to call the charge on the electron "negative" and that on a proton "positive." For that we can blame Benjamin Franklin in the eighteenth century. He didn't know about electrons and protons, but he did figure out there was a unified concept called "electric charge." When he went to arbitrarily label which substances were positively charged and which were negatively charged, he had to choose something, and the label he picked for positive charge corresponds to what we would now call "having fewer electrons than it should." So be it.

New particles and forces would be discovered over the course of the twentieth century, but in the year 1899 it wouldn't have been crazy to think that the basic picture was under control. The quantum revolution lurked just around the corner, largely unsuspected.

If you've read anything about quantum mechanics before, you've probably heard the question "Is an electron a particle, or a wave?" The answer is: "It's a wave, but when we look at (that is, measure) that wave, it looks like a particle." That's the fundamental novelty of quantum mechanics. There is only one kind of thing, the quantum wave function, but when observed under the right circumstances it appears particle-like to us.

What the World Is Made Of (Twentieth Century and Beyond)

- A quantum wave function.

It took a number of conceptual breakthroughs to go from the nineteenth-century picture of the world (classical particles and classical fields) to the twentieth-century synthesis (a single quantum wave function). The story of how particles and fields are different aspects of the same underlying thing is one of the underappreciated successes of the quest for unification in physics.

To get there, early twentieth-century physicists needed to appreciate two things: fields (like electromagnetism) can behave in particle-like ways, and particles (like electrons) can behave in wave-like ways.

The particle-like behavior of fields was appreciated first. Any particle with an electrical charge, such as an electron, creates an electric field everywhere around it, fading in magnitude as you get farther away from the charge. If we shake an electron, oscillating it up and down, the field oscillates along with it, in ripples that gradually spread out from its location. This is electromagnetic radiation, or "light" for short. Every time we heat up a material to sufficient temperature, electrons in its atoms start to shake, and the material begins to glow. This is known as *blackbody radiation*, and every object with a uniform temperature gives off a form of blackbody radiation.

Red light corresponds to slowly oscillating, low-frequency waves, while blue light is rapidly oscillating, high-frequency waves. Given what physicists knew about atoms and electrons at the turn of the century, they could calculate how much radiation a blackbody should emit at every different frequency, the so-called blackbody spectrum. Their calculations worked well for low frequencies, but became less and less accurate as they went to higher frequencies, ultimately predicting an infinite amount of radiation coming from every material body. This was later dubbed the "ultraviolet catastrophe," referring to the invisible frequencies even higher than blue or violet light.

Finally in 1900, German physicist Max Planck was able to derive a formula that fit the data exactly. The important trick was to propose a radical idea: that every time light was emitted, it came in the form of a particular amount—a "quantum"—of energy, which was related to the frequency of the light. The faster the electromagnetic field oscillates, the more energy each emission will have.

In the process, Planck was forced to posit the existence of a new fundamental parameter of nature, now known as *Planck's constant* and denoted by the letter h. The amount of energy contained in a quantum of light is proportional to its frequency, and Planck's constant is the constant of proportionality: the energy is the frequency times h. Very often it's more convenient to use a modified version $\hbar$, pronounced "h-bar," which is just Planck's original constant h divided by 2π. The appearance of Planck's constant in an expression is a signal that quantum mechanics is at work.

Planck's discovery of his constant suggested a new way of thinking about physical units, such as energy, mass, length, or time. Energy is measured in units such as ergs or joules or kilowatt-hours, while frequency is measured in units of 1/time, since frequency tells us how many times something happens in a given amount of time. To make energy proportional to frequency, Planck's constant therefore has units of energy times time. Planck himself realized that his new quantity could be combined with the other fundamental constants—G, Newton's constant of gravity, and c, the speed of light—to form universally defined measures of length, time, and so forth. The Planck length is about 10^{-33} centimeters, while the Planck time is about 10^{-43} seconds. The Planck length is a very short distance indeed, but presumably it has physical relevance, as a scale at which quantum mechanics (h), gravity (G), and relativity (c) all simultaneously matter.

Amusingly, Planck's mind immediately went to the possibility of communicating with alien civilizations. If we someday start chatting with extraterrestrial beings using interstellar radio signals, they won't know what we mean if we were to say human beings are "about two meters tall." But since they will presumably know at least as much about physics as we do, they should be aware of Planck units. This suggestion hasn't yet been put to practical use, but Planck's constant has had an immense impact elsewhere.

The idea that light is emitted in discrete quanta of energy related to its frequency is puzzling, when you think about it. From what we intuitively know about light, it might make sense if someone suggested that the amount of energy it carried depended on how *bright* it was, but not on what *color* it was. But the assumption led Planck to derive the right formula, so something about the idea seemed to be working.

It was left to Albert Einstein, in his singular way, to brush away conventional wisdom and take a dramatic leap into a new way of thinking. In 1905, Einstein suggested that light was emitted only at certain energies because it literally consisted of discrete packets, not a smooth wave. Light was particles, in other words—"photons," as they are known today. This idea, that light comes in discrete, particle-like quanta of energy, was the true birth of quantum mechanics, and was the discovery for which Einstein was awarded the Nobel Prize in 1921. (He deserved to win at least one more Nobel for the theory of relativity, but never did.) Einstein was no dummy, and he knew that this was a big deal; as he told his friend Conrad Habicht, his light quantum proposal was "very revolutionary."

Note the subtle difference between Planck's suggestion and Einstein's. Planck says that light of a fixed frequency is *emitted* in certain energy amounts, while Einstein says that's because light literally *is* discrete particles. It's the difference between saying that a certain coffee machine makes exactly one cup at a time, and saying that coffee only exists in the form of one-cup-size amounts. That might make sense when we're talking about matter particles like electrons and protons, but just a few decades earlier Maxwell had triumphantly explained that light was a wave, not a particle. Einstein's proposal was threatening to undo that triumph. Planck himself was reluctant to accept this wild new idea, but it did explain the data. In a wild new idea's search for acceptance, that's a powerful advantage to have.

◦ ◦ ◦

Meanwhile another problem was lurking over on the particle side of the ledger, where Rutherford's model explained atoms in terms of electrons orbiting nuclei.

Remember that if you shake an electron, it emits light. By "shake" we just mean accelerate in some way. An electron that does anything other than move in a straight line at a constant velocity should emit light.

From the picture of the Rutherford atom, with electrons orbiting around the nucleus, it certainly looks like those electrons are not moving in straight lines. They're moving in circles or ellipses. In a classical world, that unambiguously means that the electrons are being accelerated, and equally unambiguously that they should be giving off light. Every single atom in your body, and in the environment around you, should be glowing, if classical mechanics was right. That means the electrons should be losing energy as they emit radiation, which in turn implies that they should spiral downward into the central nucleus. Classically, electron orbits should not be stable.

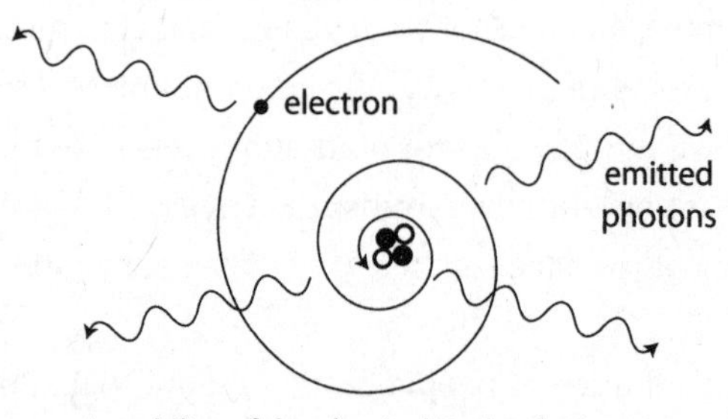

Instability of the classical Rutherford atom

Perhaps all of your atoms are giving off light, but it's just too faint to see. After all, identical logic applies to the planets in the solar system. They should be giving off gravitational waves—an accelerating mass should cause ripples in the gravitational field, just like an accelerating charge causes ripples in the electromagnetic field. And indeed they are. If there was any doubt that this happens, it was swept away in 2016, when researchers at the LIGO and Virgo gravitational-wave observatories announced the first direct detection of gravitational waves, created when black holes over a billion light years away spiraled into each other.

But the planets in the solar system are much smaller, and move more slowly, than those black holes, which were each over thirty times the mass of the sun. As a result, the emitted gravitational waves from our planetary neighbors are very weak indeed. The power emitted in gravitational waves by the orbiting Earth amounts to about 200 watts—equivalent to the output of a few lightbulbs, and completely insignificant compared to other influences such as solar radiation and tidal forces. If we pretend that the emission of gravitational waves were the only thing affecting the Earth's orbit, it would take over 10^{23} years for it to crash into the sun. So perhaps the same thing is true for atoms: maybe electron orbits aren't really stable, but they're stable enough.

This is a quantitative question, and it's not hard to plug in the numbers and see what falls out. The answer is catastrophic, because electrons should move much faster than planets and electromagnetism is a much stronger force than gravity. The amount of time it would take an electron to crash into the nucleus of its atom works out to about ten picoseconds. That's one-hundred-billionth of a second. If ordinary matter made of atoms only lasted for that long, someone would have noticed by now.

This bothered a lot of people, most notably Niels Bohr, who had briefly worked under Rutherford in 1912. In 1913, Bohr published a series of three papers, later known simply as "the trilogy," in which he put forth another of those audacious, out-of-the-blue ideas that characterized the early years of quantum theory. What if, he asked, electrons can't spiral down into atomic nuclei because electrons simply aren't allowed to be in any orbit they want, but instead have to stick to certain very specific orbits? There would be a minimum-energy orbit, another one with somewhat higher energy, and so on. But electrons weren't allowed to go any closer to the nucleus than the lowest orbit, and they weren't allowed to be in between the orbits. The allowed orbits were quantized.

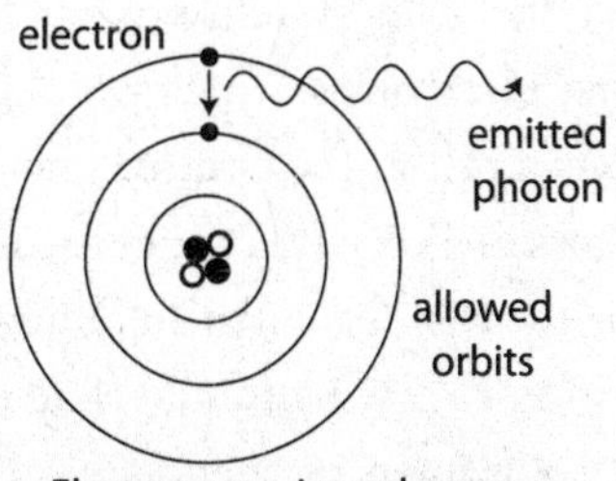

Electrons can jump between allowed orbits in the Bohr atom

Bohr's proposal wasn't quite as outlandish as it might seem at first. Physicists had studied how light interacted with different elements in their gaseous form—hydrogen, nitrogen, oxygen, and so forth. They found that if you shined light through a cold gas, some of it would be absorbed; likewise, if you passed electrical current through a tube of gas, the gas would start glowing (the principle behind fluorescent lights still used today). But they only emitted and absorbed certain very specific frequencies of light, letting other colors pass right through. Hydrogen, the simplest element with just a single proton and a single electron, in particular had a very regular pattern of emission and absorption frequencies.

For a classical Rutherford atom, that would make no sense at all. But in Bohr's model, where only certain electron orbits were allowed, there was an immediate explanation. Even though electrons couldn't linger in between the allowed orbits, they could *jump* from one to another. An electron could fall from a higher-energy orbit to a lower-energy one by emitting light with just the right energy to compensate, or it could leap upward in energy by absorbing an appropriate amount of energy from ambient light. Because the orbits themselves were quantized, we should only see specific energies of light interacting with the electrons. Together with Planck's idea that the frequency of light is related to its energy, this explained why physicists saw only certain frequencies being emitted or absorbed.

By comparing his predictions to the observed emission of light by hydrogen, Bohr was able to not simply posit that only some electron orbits were allowed, but calculate which ones they were. Any orbiting particle has a quantity called the *angular momentum*, which is easy to calculate—it's just the mass of the particle, times its velocity, times its distance from the center of the orbit. Bohr proposed that an allowed electron orbit was one whose angular momentum was a multiple of a particular fundamental constant. And when he compared the energy that electrons should emit when jumping between orbits to what was actually seen in light emitted from hydrogen gas, he could figure out what that constant needed to be in order to fit the data. The answer was Planck's constant, h. Or more specifically, the modified h-bar version, $\hbar = h/2\pi$.

That's the kind of thing that makes you think you're on the right track. Bohr was trying to account for the behavior of electrons in atoms, and he posited an ad hoc rule according to which they could only move along certain quantized orbits, and in order to fit the data his rule ended up requiring a new constant of nature, and that new constant was the same as the new constant that Planck was forced to invent when he was trying to account for the behavior of photons. All of this might seem ramshackle and a bit sketchy, but taken together it appeared as if something profound was happening in the realm of atoms and particles, something that didn't fit comfortably with the sacred rules of classical mechanics. The ideas of this period are now sometimes described under the rubric of "the old quantum theory," as opposed to "the new quantum theory" of Heisenberg and Schrödinger that came along in the late 1920s.

∘ ∘ ∘

As provocative and provisionally successful as the old quantum theory was, nobody was really happy with it. Planck and Einstein's idea of light quanta helped make sense of a number of experimental results, but was hard to reconcile with the enormous success of Maxwell's theory of light as electromagnetic waves. Bohr's idea of quantized electron orbits helped make sense of the light emitted and absorbed by hydrogen, but seemed to be pulled out of a hat, and didn't really work for elements other than hydrogen. Even before the "old quantum theory" was given that name, it seemed clear that these were just hints at something much deeper going on.

One of the least satisfying features of Bohr's model was the suggestion that electrons could "jump" from one orbit to another. If a low-energy electron absorbed light with a certain amount of energy, it makes sense that it would have to jump up to another orbit with just the right amount of additional energy. But when an electron in a high-energy orbit emitted light to jump down, it seemed to have a choice about exactly how far down to go, which lower orbit to end up in. What made that choice? Rutherford himself worried about this in a letter to Bohr:

> There appears to me one grave difficulty in your hypothesis, which I have no doubt you fully realize, namely, how does an electron decide what frequency it is going to vibrate at when it passes from one stationary state to the other? It seems to me that you would have to assume that the electron knows beforehand where it is going to stop.

This business about electrons "deciding" where to go foreshadowed a much more drastic break with the paradigm of classical physics than physicists in 1913 were prepared to contemplate. In Newtonian mechanics one could imagine a Laplace demon that could predict, at least in principle, the entire future history of the world from its present state. At this point in the development of quantum mechanics, nobody was really confronting the prospect that this picture would have to be completely discarded.

It took more than ten years for a more complete framework, the "new quantum theory," to finally come on the scene. In fact, two competing ideas were proposed at the time, matrix mechanics and wave mechanics, before they were ultimately shown to be mathematically equivalent versions of the same thing, which can now simply be called quantum mechanics.

Matrix mechanics was formulated initially by Werner Heisenberg, who had worked with Niels Bohr in Copenhagen. These two men, along with their collaborator Wolfgang Pauli, are responsible for the Copenhagen interpretation of quantum mechanics, though who exactly believed what is a topic of ongoing historical and philosophical debate.

Heisenberg's approach in 1926, reflecting the boldness of a younger generation coming on the scene, was to put aside questions of what was really happening in a quantum system, and to focus exclusively on explaining what was observed by experimenters. Bohr had posited quantized electron orbits without explaining why some orbits were allowed and others were not. Heisenberg dispensed with orbits entirely. Forget about what the electron is doing; ask only what you can observe about it. In classical mechanics, an electron would be characterized by position and momentum. Heisenberg kept those words, but instead of thinking of them as quantities that exist whether we are looking at them or not, he thought of them as possible outcomes of measurements. For Heisenberg, the unpredictable jumps that had bothered Rutherford and others became a central part of the best way of talking about the quantum world.

Heisenberg was only twenty-four years old when he first formulated matrix mechanics. He was clearly a prodigy, but far from an established figure in the field, and wouldn't obtain a permanent academic position until a year later. In a letter to Max Born, another of his mentors, Heisenberg fretted that he "had written a crazy paper and did not dare to send it in for publication." But in a collaboration with Born and the even younger physicist Pascual Jordan, they were able to put matrix mechanics on a clear and mathematically sound footing.

It would have been natural for Heisenberg, Born, and Jordan to share the Nobel Prize for the development of matrix mechanics, and indeed Einstein nominated them for the award. But it was Heisenberg alone who was honored by the Nobel committee in 1932. It has been speculated that Jordan's inclusion would have been problematic, as he became known for aggressive right-wing political rhetoric, ultimately becoming a member of the Nazi Party and joining a *Sturmabteilung* (Storm trooper) unit. At the same time, however, he was considered unreliable by his fellow Nazis, due to his support for Einstein and other Jewish scientists. In the end, Jordan never won the prize. Born was also left off the prize for matrix mechanics, but that omission was made up for when he was awarded a separate Nobel in 1954 for his formulation of the probability rule. That was the last time a Nobel Prize has been awarded for work in the foundations of quantum mechanics.

After the onset of World War II, Heisenberg led a German government program to develop nuclear weapons. What Heisenberg actually thought about the Nazis, and whether he truly tried as hard as possible to push the weapons program forward, are matters of some historical dispute. It seems that, like a number of other Germans, Heisenberg was not fond of the Nazi Party, but preferred a German victory in the conflict to the prospect of being run over by the Soviets. There is no evidence that he actively worked to sabotage the nuclear bomb program, but it is clear that his team made very little progress. In part that must be attributed to the fact that so many brilliant Jewish physicists had fled Germany as the Nazis took power.

◦ ◦ ◦

As impressive as matrix mechanics was, it suffered from a severe marketing flaw: the mathematical formalism was highly abstract and difficult to understand. Einstein's reaction to the theory was typical: "A veritable sorcerer's calculation. This is sufficiently ingenious and protected by its great complexity, to be immune to any proof of its falsity." (This from the guy who had proposed describing spacetime in terms of non-Euclidean geometry.) Wave mechanics, developed immediately thereafter by Erwin Schrödinger, was a version of quantum theory that used concepts with which physicists were already very familiar, which greatly helped accelerate acceptance of the new paradigm.

Physicists had studied waves for a long time, and with Maxwell's formulation of electromagnetism as a theory of fields, they had become adept at thinking about them. The earliest intimations of quantum mechanics, from Planck and Einstein, had been away from waves and toward particles. But Bohr's atom suggested that even particles weren't what they seemed to be.

In 1924, the young French physicist Louis de Broglie was thinking about Einstein's light quanta. At this point the relationship between photons and classical electromagnetic waves was still murky. An obvious thing to contemplate was that light consisted of both a particle and a wave: particle-like photons could be carried along by the well-known electromagnetic waves. And if that's true, there's no reason we couldn't imagine the same thing going on with electrons—maybe there is something wave-like that carries along the electron particles. That's exactly what de Broglie suggested in his 1924 doctoral thesis, proposing a relationship between the momentum and wavelength of these "matter waves" that was analogous to Planck's formula for light, with larger momenta corresponding to shorter wavelengths.

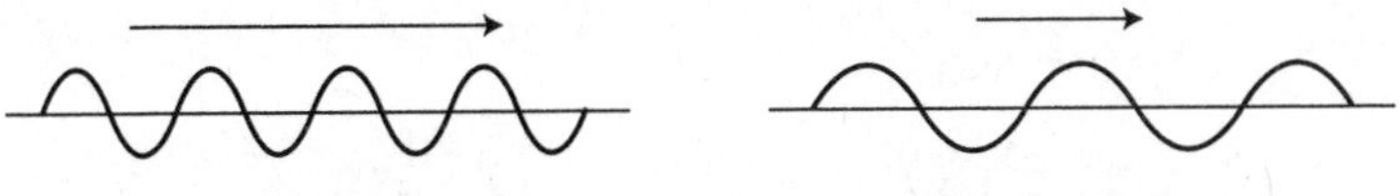

short wavelength =
high energy, large momentum

long wavelength =
low energy, small momentum

Like many suggestions at the time, de Broglie's hypothesis may have seemed a little ad hoc, but its implications were far-reaching. In particular, it was natural to ask what the implications of matter waves might be for electrons orbiting around a nucleus. A remarkable answer suggested itself: for the wave to settle down into a stationary configuration, its wavelength had to be an exact multiple of the circumference of a corresponding orbit. Bohr's quantized orbits could be derived rather than simply postulated, simply by associating waves with the electron particles surrounding the nucleus.

Consider a string with its ends held fixed, such as on a guitar or violin. Even though any one point can move up or down as it likes, the overall behavior of the string is constrained by being tied down at either end. As a result, the string only vibrates at certain special wavelengths, or combinations thereof; that's why the strings on musical instruments emit clear notes rather than an indistinct noise. These special vibrations are called the *modes* of the string. The essentially "quantum" nature of the subatomic world, in this picture, comes about not because reality is actually subdivided into distinct chunks but because there are natural vibrational modes for the waves out of which physical systems are made.

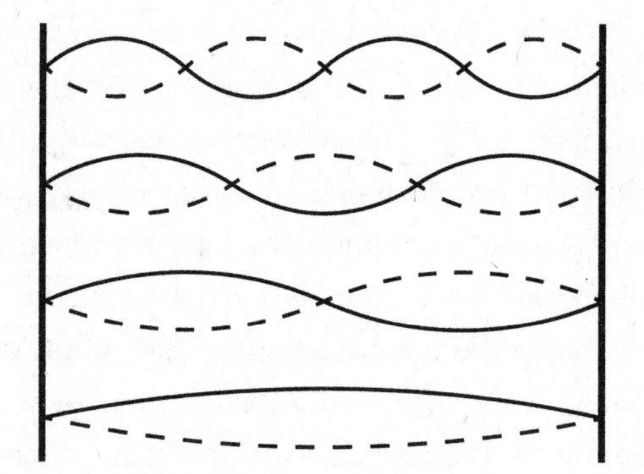

Allowed wavelengths (modes)
of a string with ends tied down

The word "quantum," referring to some definite amount of stuff, can give the impression that quantum mechanics describes a world that is fundamentally discrete and pixelated, like when you zoom in closely on a computer monitor or TV screen. It's actually the opposite; quantum mechanics describes the world as a smooth wave function. But in the right circumstances, where individual parts of the wave function are tied down in a certain way, the wave takes the form of a combination of distinct vibrational modes. When we observe such a system, we see those discrete possibilities. That's true for orbits of electrons, and it will also explain why quantum fields look like sets of individual particles. In quantum mechanics, the world is fundamentally wavy; its apparent quantum discreteness comes from the particular way those waves are able to vibrate.

De Broglie's ideas were intriguing, but they fell short of providing a comprehensive theory. That was left to Erwin Schrödinger, who in 1926 put forth a dynamical understanding of wave functions, including the equation they obey, later named after him. Revolutions in physics are generally a young person's game, and quantum mechanics was no different, but Schrödinger bucked the trend. Among the leaders of the discussions at Solvay in 1927, Einstein at forty-eight years old, Bohr at forty-two, and Born at forty-four were the grand old men. Heisenberg was twenty-five, Pauli twenty-seven, and Dirac twenty-five. Schrödinger, at the ripe old age of thirty-eight, was looked upon as someone suspiciously long in the tooth to appear on the scene with radical new ideas like this.

Note the shift here from de Broglie's "matter waves" to Schrödinger's "wave function." Though Schrödinger was heavily influenced by de Broglie's work, his concept went quite a bit further, and deserves a distinct name. Most obviously, the value of a matter wave at any one point was some real number, while the amplitudes described by wave functions are complex numbers—the sum of a real number and an imaginary one.

More important, the original idea was that each kind of particle would be associated with a matter wave. That's not how Schrödinger's wave function works; you have just one function that depends on all the particles in the universe. It's that simple shift that leads to the world-altering phenomenon of quantum entanglement.

○ ○ ○

What made Schrödinger's ideas an instant hit was the equation he proposed, which governs how wave functions change with time. To a physicist, a good equation makes all the difference. It elevates a pretty-sounding idea ("particles have wave-like properties") to a rigorous, unforgiving framework. Unforgiving might sound like a bad quality in a person, but it's just what you want in a scientific theory. It's the feature that lets you make precise predictions. When we say that quantum textbooks spend a lot of time having students solve equations, it's mostly the Schrödinger equation we have in mind.

Schrödinger's equation is what a quantum version of Laplace's demon would be solving as it predicted the future of the universe. And while the original form in which Schrödinger wrote down his equation was meant for systems of individual particles, it's actually a very general idea that applies equally well to spins, fields, superstrings, or any other system you might want to describe using quantum mechanics.

Unlike matrix mechanics, which was expressed in terms of mathematical concepts most physicists at the time had never been exposed to, Schrödinger's wave equation was not all that different in form from Maxwell's electromagnetic equations that adorn T-shirts worn by physics students to this day. You could visualize a wave function, or at least you might convince yourself that you could. The community wasn't sure what to make of Heisenberg, but they were ready for Schrödinger. The Copenhagen crew—especially the youngsters, Heisenberg and Pauli—didn't react graciously to the competing ideas from an undistinguished old man in Zürich. But before too long they were thinking in terms of wave functions, just like everyone else.

Schrödinger's equation involves unfamiliar symbols, but its basic message is not hard to understand. De Broglie had suggested that the momentum of a wave goes up as its wavelength goes down. Schrödinger proposed a similar thing, but for energy and time: the rate at which the wave function is changing is proportional to how much energy it has. Here is the celebrated equation in its most general form:

$$\frac{\partial \Psi}{\partial t} = \frac{1}{i\hbar} H\Psi$$

We don't need the details here, but it's nice to see the real way that physicists think of an equation like this. There's some math involved, but ultimately it's just a translation into symbols of the idea we wrote down in words.

Ψ (the Greek letter Psi) is the wave function. The left-hand side is the rate at which the wave function is changing over time. On the right-hand side we have a proportionality constant involving Planck's constant $\hbar$, the fundamental unit of quantum mechanics, and i, the square root of minus one. The wave function Ψ is acted on by something called the *Hamiltonian*, or H. Think of the Hamiltonian as an inquisitor who asks the following question: "How much energy do you have?" The concept was invented in 1833 by Irish mathematician William Rowan Hamilton, as a way to reformulate the laws of motion of a classical system, long before it gained a central role in quantum mechanics.

When physicists start modeling different physical systems, the first thing they try to do is work out a mathematical expression for the Hamiltonian of that system. The standard way of figuring out the Hamiltonian of something like a collection of particles is to start with the energies of the particles themselves, and then add in additional contributions describing how the particles interact with each other. Maybe they bump off each other like billiard balls, or perhaps they exert a mutual gravitational interaction. Each such possibility suggests a particular kind of Hamiltonian. And if you know the Hamiltonian, you know everything; it's a compact way of capturing all the dynamics of a physical system.

If a quantum wave function describes a system with some definite value of the energy, the Hamiltonian simply equals that value, and the Schrödinger equation implies that the system just keeps doing the same thing, maintaining a fixed energy. More often, since wave functions are superpositions of different possibilities, the system will be a combination of multiple energies. In that case the Hamiltonian captures a bit of all of them. The bottom line is that the right-hand side of Schrödinger's equation is a way of characterizing how much energy is carried by each of the contributions to a wave function in a quantum superposition; high-energy components evolve quickly, low-energy ones evolve more slowly.

What really matters is that there is some specific deterministic equation. Once you have that, the world is your playground.

○ ○ ○

Wave mechanics made a huge splash, and before too long Schrödinger, English physicist Paul Dirac, and others demonstrated that it was essentially equivalent to matrix mechanics, leaving us with a unified theory of the quantum world. Still, all was not peaches and cream. Physicists were left with the question that we are still struggling with today: What *is* the wave function, really? What physical thing does it represent, if any?

In de Broglie's view, his matter waves served to guide particles around, not to replace them entirely. (He later developed this idea into pilot-wave theory, which remains a viable approach to quantum foundations today, although it is not popular among working physicists.) Schrödinger, by contrast, wanted to do away with fundamental particles entirely. His original hope was that his equation would describe localized packets of vibrations, confined to a relatively small region of space, so that each packet would appear particle-like to a macroscopic observer. The wave function could be thought of as representing the density of mass in space.

Alas, Schrödinger's aspirations were undone by his own equation. If we start with a wave function describing a single particle approximately localized in some empty region of space, the Schrödinger equation is clear about what happens next: it quickly spreads out all over the place. Left to their own devices, Schrödinger's wave functions don't look particle-like at all.*

It was left to Max Born, one of Heisenberg's collaborators on matrix mechanics, to provide the final missing piece: we should think about the wave function as a way of calculating the probability of seeing a particle in any given position when we look for it. In particular, we

* I've emphasized that there is only one wave function, the wave function of the universe, but the alert reader will notice that I often talk about "the wave function of a particle." This latter construction is perfectly okay if—and only if—the particle is unentangled from the rest of the universe. Happily, that is often the case, but in general we have to keep our wits about us.

should take both the real and imaginary parts of the complex-valued amplitude, square them both individually, and add the two numbers together. The result is the probability of observing the corresponding outcome. (The suggestion that it's the amplitude squared, rather than the amplitude itself, appears in a footnote added at the last minute to Born's 1926 paper.) And after we observe it, the wave function collapses to be localized at the place where we saw the particle.

You know who didn't like the probability interpretation of the Schrödinger equation? Schrödinger himself. His goal, like Einstein's, was to provide a definite mechanistic underpinning for quantum phenomena, not just to create a tool that could be used to calculate probabilities. "I don't like it, and I'm sorry I ever had anything to do with it," he later groused. The point of the famous Schrödinger's Cat thought experiment, in which the wave function of a cat evolves (via the Schrödinger equation) into a superposition of "alive" and "dead," was not to make people say, "Wow, quantum mechanics is really mysterious." It was to make people say, "Wow, this can't possibly be correct." But to the best of our current knowledge, it is.

◦ ◦ ◦

A lot of intellectual action was packed into the first three decades of the twentieth century. Over the course of the 1800s, physicists had put together a promising picture of the nature of matter and forces. Matter was made of particles, and forces were carried by fields, all under the umbrella of classical mechanics. But confrontation with experimental data forced them to think beyond this paradigm. In order to explain radiation from hot objects, Planck suggested that light was emitted in discrete amounts of energy, and Einstein pushed this further by suggesting that light actually came in the form of particle-like quanta. Meanwhile, the fact that atoms are stable and the observation of how light was emitted from gases inspired Bohr to suggest that electrons could only move along certain allowed orbits, with occasional jumps between them. Heisenberg, Born, and Jordan elaborated this story of probabilistic jumps into a full theory, matrix mechanics. From another angle, de Broglie pointed out that if we think of matter particles such as electrons as actually being waves, we can derive Bohr's quantized orbits rather than postulating them. Schrödinger developed this suggestion into a full-blown quantum theory of its own, and it was ultimately demonstrated that wave mechanics and matrix mechanics were equivalent ways of saying the same thing. Despite initial hopes that wave mechanics could explain away the apparent need for probabilities as a fundamental part of the theory, Born showed that the right way to think about Schrödinger's wave function was as something that you square to get the probability of a measurement outcome.

Whew. That's quite a journey, taken in a remarkably short period of time, from Planck's observation in 1900 to the Solvay Conference in 1927, when the new quantum mechanics was fleshed out once and for all. It's to the enormous credit of the physicists of the early twentieth century that they were willing to face up to the demands of the experimental data, and in doing so to completely upend the fantastically successful Newtonian view of the classical world.

They were less successful, however, at coming to grips with the implications of what they had wrought.

What Cannot Be Known, Because It Does Not Exist

Uncertainty and Complementarity

A police officer pulls over Werner Heisenberg for speeding. "Do you know how fast you were going?" asks the cop. "No," Heisenberg replies, "but I know exactly where I am!"

I think we can all agree that physics jokes are the funniest jokes there are. They are less good at accurately conveying physics. This particular chestnut rests on familiarity with the famous Heisenberg uncertainty principle, often explained as saying that we cannot simultaneously know both the position and the velocity of any object. But the reality is deeper than that.

It's not that we can't know position and momentum, it's that they don't even exist at the same time. Only under extremely special circumstances can an object be said to have a location—when its wave function is entirely concentrated on one point in space, and zero everywhere else—and similarly for velocity. And when one of the two is precisely defined, the other could be literally anything, were we to measure it.

More often, the wave function includes a spread of possibilities for both quantities, so neither has a definite value.

Back in the 1920s, all this was less clear. It was still natural to think that the probabilistic nature of quantum mechanics simply indicated that it was an incomplete theory, and that there was a more deterministic, classical-sounding picture waiting to be developed. Wave functions, in other words, might be a way of characterizing our ignorance of what was really going on, rather than being the total truth about what is going on, as we're advocating here. One of the first things people did when learning about the uncertainty principle was to try to find loopholes in it. They failed, but in doing so we learned a lot about how quantum reality is fundamentally different from the classical world we had been used to.

The absence of definite quantities at the heart of reality that map more or less straightforwardly onto what we can eventually observe is one of the deep features of quantum mechanics that can be hard to accept upon first encounter. There are quantities that are not merely unknown but do not even exist, even though we can seemingly measure them.

Quantum mechanics forces us to confront this yawning chasm between what we see and what really is. In this chapter we'll see how that gap manifests itself in the uncertainty principle, and in the next chapter we'll see it again more forcefully in the phenomenon of entanglement.

○ ○ ○

The uncertainty principle owes its existence to the fact that the relationship between position and momentum (mass times velocity) is fundamentally different in quantum mechanics from what it was in classical mechanics. Classically, we can imagine measuring the momentum of a particle by tracking its position over time, and seeing how fast it moves. But if all we have access to is a single moment, position and momentum are completely independent from each other. If I tell you that a particle has a certain position at one instant, and I tell you nothing else, you have no idea what its speed is, and vice versa.

Physicists refer to the different numbers we use to specify something as that system's "degrees of freedom." In Newtonian mechanics, to tell me the complete state of a bunch of particles, you have to tell me the position and momentum of every one of them, so the degrees of freedom are the positions and the momenta. Acceleration is not a degree of freedom, since it can be calculated once we know the forces acting on the system. The essence of a degree of freedom is that it doesn't depend on anything else.

When we switch to quantum mechanics and start thinking about Schrödinger's wave functions, things become a little different. To make a wave function for a single particle, think of every location where the particle could possibly be found, were we to observe it. Then to each location assign an amplitude, a complex number with the property that the square of each number is the probability of finding the particle there. There is a constraint that the squares of all these numbers add up to precisely one, since the total probability that the particle is found somewhere must equal one. (Sometimes we speak of probabilities in terms of percentages, which are numerically 100 times the actual probability; a 20 percent chance is the same as a 0.2 probability.)

Notice we didn't mention "velocity" or "momentum" there. That's because we don't have to separately specify the momentum in quantum mechanics, as we did in classical mechanics. The probability of measuring any particular velocity is completely determined by the wave function for all the possible positions. Velocity is not a separate degree of freedom, independent of position. The basic reason why is that the wave function is, you know, a wave. Unlike for a classical particle, we don't have a single position and a single momentum, we have a function of all possible positions, and that function typically oscillates up and down. The rate of those oscillations determines what we're likely to see if we were to measure the velocity or momentum.

Consider a simple sine wave, oscillating up and down in a regular pattern throughout space. Plug such a wave function into the Schrödinger equation and ask how it will evolve. We find that a sine wave has a definite momentum, with shorter wavelengths corresponding to faster velocity. But a sine wave has no definite position; on the contrary, it's spread out everywhere. And a more typical shape, which is neither localized at one point nor spread out in a perfect sine wave of fixed wavelength, won't correspond to either a definite position or a definite momentum, but some mixture of each.

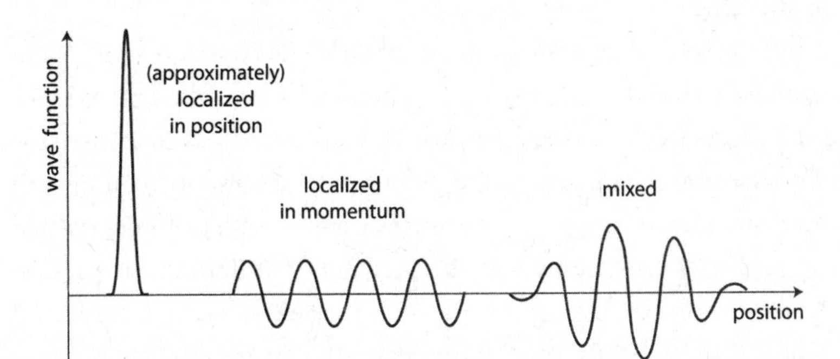

We see the basic dilemma. If we try to localize a wave function in space, its momentum becomes more and more spread out, and if we try to limit it to one fixed wavelength (and therefore momentum) it becomes more spread out in position. That's the uncertainty principle. It's not that we can't *know* both quantities at the same time; it's just a fact about how wave functions work that if position is concentrated near some location, momentum is completely undetermined, and vice versa. The old-fashioned classical properties called *position* and *momentum* aren't quantities with actual values, they're possible measurement outcomes.

People sometimes refer to the uncertainty principle in everyday contexts, outside of the equation-filled language of physics texts. So it's important to emphasize what the principle does *not* say. It's not an assertion that "everything is uncertain." Either position or momentum could be certain in an appropriate quantum state; they just can't be certain at the same time.

And the uncertainty principle doesn't say we necessarily disturb a system when we measure it. If a particle has a definite momentum, we can go ahead and measure that without changing it at all. The point is that there are no states for which both position and momentum are simultaneously definite. The uncertainty principle is a statement about the nature of quantum states and their relationship to observable quantities, not a statement about the physical act of measurement.

Finally, the principle is not a statement about limitations on our knowledge of the system. We can know the quantum state exactly, and that's all there is to know about it; we still can't predict the results of all possible future observations with perfect certainty. The idea that "there's something we don't know," given a certain wave function, is an outdated relic of our intuitive insistence that what we observe is what really exists. Quantum mechanics teaches us otherwise.

○ ○ ○

You'll sometimes hear the idea, provoked by the uncertainty principle, that quantum mechanics *violates logic itself.* That's silly. Logic deduces theorems from axioms, and the resulting theorems are simply true. The axioms may or may not apply to any given physical situation. Pythagoras's theorem—the square of the hypotenuse of a right triangle equals the sum of the squares of the other two sides—is correct as a formal deduction from the axioms of Euclidean geometry, even though those axioms do not hold if we're talking about curved surfaces rather than a flat tabletop.

The idea that quantum mechanics violates logic lives in the same neighborhood of the idea that atoms are mostly empty space (a bad neighborhood). Both notions stem from a deep conviction that, despite everything we've learned, particles are really points with some position and momentum, rather than being wave functions that are spread out.

Consider a particle in a box, where we've drawn a line dividing the box into left and right sides. It has some wave function that is spread throughout the box. Let proposition P be "the particle is on the left side of the box," and proposition Q be "the particle is on the right side of the box." We might be tempted to say that both of these propositions are false, since the wave function stretches over both sides of the box. But the proposition "P or Q" has to be true, since the particle is in the box. In classical logic, we can't have both P and Q be false but "P or Q" be true. So something fishy is going on.

What's fishy is neither logic nor quantum mechanics but our casual disregard for the nature of quantum states when assigning truth values to the statements P and Q. These statements are neither true nor false; they're just ill defined. There is no such thing as "the side of the box the particle is on." If the wave function were concentrated entirely on one side of the box and exactly vanished on the other, we could get away with assigning truth values to P and Q; but in that case one would be true and the other would be false, and classical logic would be fine.

Despite the fact that classical logic is perfectly valid whenever it is properly applied, quantum mechanics has inspired more general approaches known as *quantum logic*, pioneered by John von Neumann and his collaborator Garrett Birkhoff. By starting with slightly different logical axioms from the standard ones, we can derive a set of rules obeyed by the probabilities implied by the Born rule in quantum mechanics. Quantum logic in this sense is both interesting and useful, but its existence does not invalidate the correctness of ordinary logic in appropriate circumstances.

◦ ◦ ◦

Niels Bohr, in an attempt to capture what makes quantum theory so unique, proposed the concept of *complementarity*. The idea is that there can be more than one way of looking at a quantum system, each of them equally valid, but with the property that you can't employ them simultaneously. We can describe the wave function of a particle in terms of either position or momentum, but not both at the same time. Similarly, we can think of electrons as exhibiting either particle-like or wave-like properties, just not at the same time.

Nowhere is this feature made more evident than in the famous double-slit experiment. This experiment wasn't actually performed until the 1970s, long after it was proposed. It wasn't one of those surprising experimental results that theorists had to invent a new way of thinking in order to understand, but rather a thought experiment (suggested in its original form by Einstein during his debates with Bohr, and later popularized by Richard Feynman in his lectures to Caltech undergraduates) meant to show the dramatic implications of quantum theory.

The idea of the experiment is to home in on the distinction between particles and waves. We start with a source of classical particles (maybe a pellet gun that tends to spray in somewhat unpredictable directions), shoot them through a single thin slit, then detect them at a screen on the other side of the slit. Mostly the particles will pass right through, with perhaps very slight deviations if they bump up against the sides of the slit. So what we see at the detector is a pattern of individual points where we detect the particles, arranged in more or less a slit-like pattern.

We could also do the same thing with waves, for example, by placing the slit in a tub of water and creating waves that pass through it. When the waves pass through, they spread out in a semicircular pattern before eventually reaching the screen. Of course, we don't observe particle-like points when the water wave hits the screen, but let's imagine we have a special screen that lights up with a brightness that depends on the amplitude the waves reach at any particular point. They will be brightest at the point of the screen that is closest to the slit, and gradually fade as we get farther away.

Now let's do the same thing, but with two slits in the way rather than just one. The particle case isn't that much different; as long as our

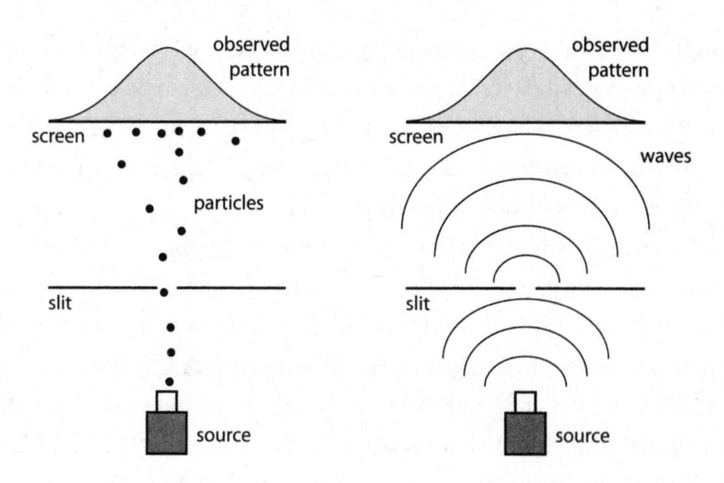

source of particles is sufficiently random that particles pass through both slits, what we'll see on the other side is two lines of points, one for each slit (or one thick line, if the slits themselves are sufficiently close together). But the wave case is altered in an interesting way. Waves can oscillate downward as well as upward, and two waves oscillating in opposite directions will cancel each other out—a phenomenon known as *interference*. So the waves pass through both slits at once, emanating outward in semicircles, but then set up an interference pattern on the other side. As a result, if we observe the amplitude of the resultant wave at the final screen, we don't simply see two bright lines; rather, there will be a bright line in the middle (closest to both slits), with alternating dark/bright regions that gradually fade to either side.

So far, that's the classical world we know and love, where particles and waves are different things and everyone can easily distinguish between them. Now let's replace our pellet gun or wave machine with a source of electrons, in all their quantum-mechanical glory. There are several twists on this setup, each with provocative consequences.

First consider just a single slit. In this case the electrons behave just as if they were classical particles. They pass through the slit, then are detected by the screen on the other side, each electron leaving a single

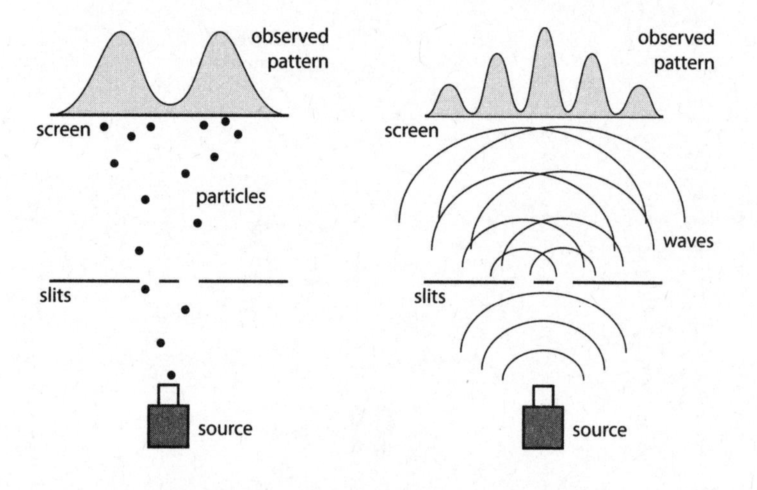

particle-like mark. If we let numerous electrons through, their marks are scattered around a central line in the image of the slit that they passed through. Nothing funny yet.

Now let's introduce two slits. (The slits have to be very close together for this to work, which is one reason it took so long for the experiment to actually be carried out.) Once again, electrons pass through the slits and leave individual marks on the screen on the other side. However, their marks do *not* clump into two lines, as the classical pellets did. Rather, they form a series of lines: a high-density one in the middle, surrounded by parallel lines with gradually fewer marks, each separated by dark regions with almost no marks at all.

In other words, electrons going through two slits leave what is unmistakably an interference pattern, just like waves do, even though they hit the screen with individual marks just like particles. This phenomenon has launched a thousand unhelpful discussions about whether electrons are "really" particles or waves, or are sometimes particle-like and other times wave-like. One way or another, it's indisputable that something went through both slits as the electrons traveled to the screen.

At this point this is no surprise to us. The electrons passing through

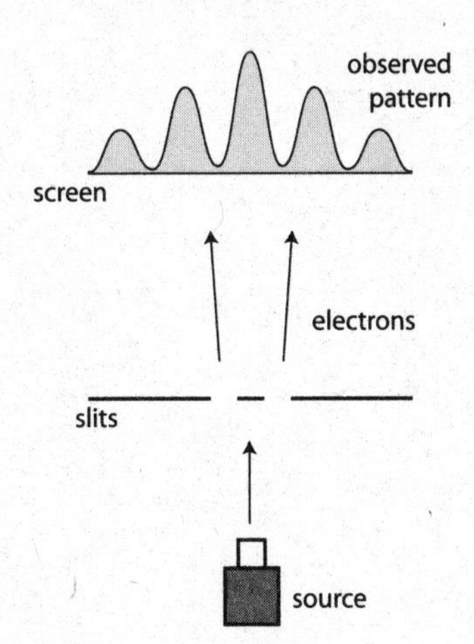

the slits are described by a wave function, which just like our classical wave will go through both slits and oscillate up and down, and therefore it makes sense that we see interference patterns. Then when they hit the screen they are being observed, and it's at that point they appear to us as particles.

Let's introduce one additional wrinkle. Imagine that we set up little detectors at each slit, so we can tell whether an electron goes through it. That will settle this crazy idea that an electron can travel through two slits once and for all.

You should be able to figure out what we see. The detectors don't measure half of an electron going through each of the two slits; they measure a full electron going through one, and nothing through the other, every time. That's because the detector acts as a measuring device, and when we measure electrons we see particles.

But that's not the only consequence of looking at the electron as it passes through the slits. At the screen, on the other side of the slits, the interference pattern disappears, and we are back to seeing two bands of marks made by the detected electrons, one for each slit. With the detectors doing their job, the wave function collapses as the electron goes through the slits, so we don't see interference from a wave passing

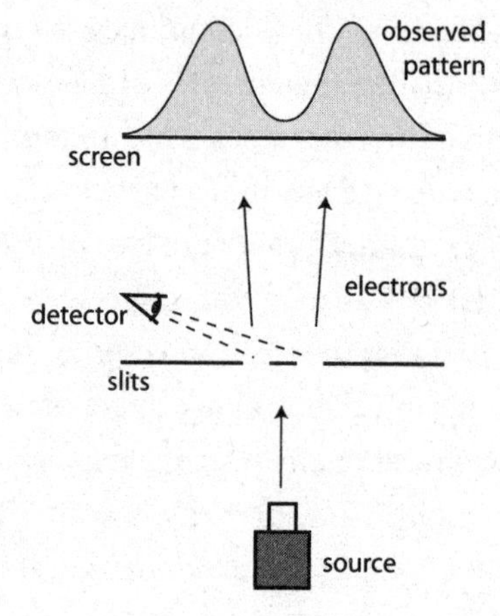

through both slits at once. When we're looking at them, electrons behave like particles.

The double-slit experiment makes it difficult to cling to the belief that the electron is just a single classical point, and the wave function simply represents our ignorance about where that point is. Ignorance doesn't cause interference patterns. There is something real about the wave function.

∘ ∘ ∘

Wave functions may be real, but they're undeniably abstract, and once we start considering more than one particle at a time they become hard to visualize. As we move forward with increasingly subtle examples of quantum phenomena in action, it will be very helpful to have a simple, readily graspable example we can refer to over and over. The *spin* of a particle—a degree of freedom in addition to its position or momentum—is just what we're looking for. We have to think a bit about what spin means within quantum mechanics, but once we do, it will make our lives much easier.

The notion of spin itself isn't hard to grasp: it's just rotation around an axis, as the Earth does every day or a pirouetting ballet dancer does on their tiptoes. But just like the energies of an electron orbiting an atomic nucleus, in quantum mechanics there are only certain discrete results we can obtain when we measure a particle's spin.

For an electron, for example, there are two possible measurement outcomes for spin. First pick an axis with respect to which we measure the spin. We always find that the electron is spinning either clockwise or counterclockwise when we look along that axis, and always at the same rate. These are conventionally referred to as "spin-up" and "spin-down." Think of the "right-hand rule": if you wrap the fingers of your right hand in the direction of rotation, your thumb will be pointing along the appropriate up/down axis.

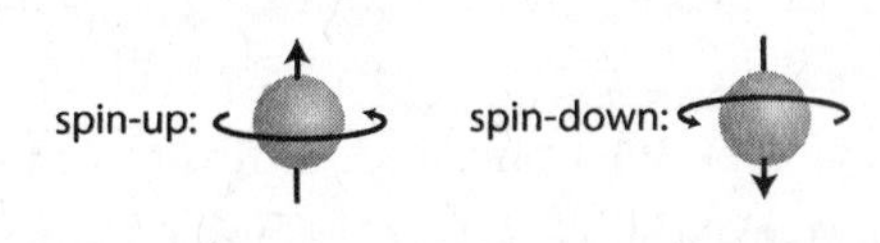

A spinning electron is a tiny magnet, with north and south magnetic poles, much like the Earth; the spin axis points toward the north pole. One way of measuring the spin of a particular electron is to shoot it through a magnetic field, which will deflect the electron by a bit depending on how its spin is oriented. (As a technicality, the magnetic field has to be focused in the right way—spread out on one side, pinched tightly on the other—for this to work.)

If I told you that the electron had a certain total spin, you might make the following prediction for such an experiment: the electron would be deflected up if its spin axis were aligned with the external field, deflected down if its spin were aligned in the opposite direction, and deflected at some intermediate angle if its spin were somewhere in between. But that's not what we see.

This experiment was first performed in 1922, by German physicists Otto Stern (an assistant to Max Born) and Walter Gerlach, before the idea of spin had been explicitly spelled out. What they saw was remarkable. Electrons are indeed deflected by passing through the magnetic field, but they either go up, or they go down; nothing in between. If we rotate the magnetic field, the electrons are still deflected in the direction of the field they pass through, either along or against it, but no intermediate values. The measured spin, like the energy of an electron orbiting an atomic nucleus, appears to be quantized.

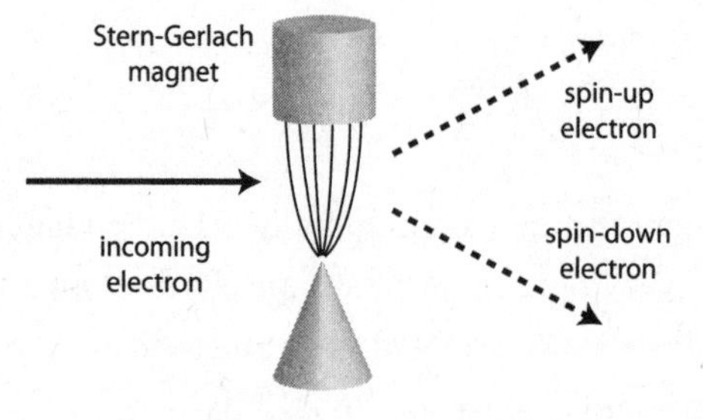

That seems surprising. Even if we've acclimated ourselves to the idea that the energy of an electron orbiting a nucleus only comes in certain quantized values, at least that energy seems like an objective property of the electron. But this thing we call the "spin" of the electron seems to give us different answers depending on how we measure it. No matter what particular direction we measure the spin along, there are only two possible outcomes we can obtain.

To make sure we haven't lost our minds, let's be clever and run the electron through two magnets in a row. Remember that the rules of textbook quantum mechanics tell us that if we get a certain measurement outcome, then measure the same system immediately again, we will always get the same answer. And indeed that's what happens; if an electron is deflected upward by one magnet (and is therefore spin-up), it will always be deflected upward by a following magnet oriented in the same way.

What if we rotate one of the magnets by 90 degrees? So we're splitting an initial beam of electrons into spin-up and spin-down as measured by a vertically oriented magnet, then taking the spin-up electrons

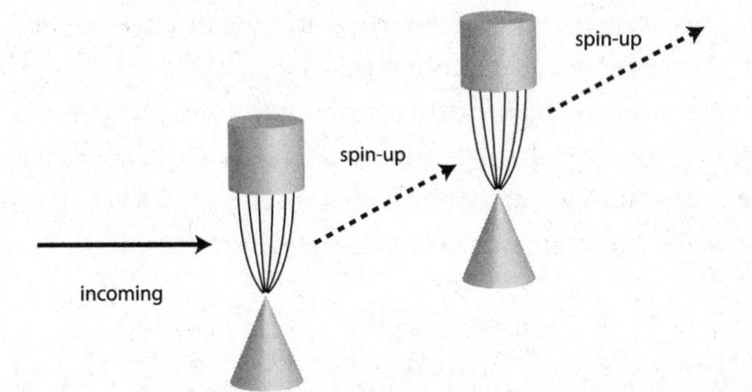

and passing them through a horizontally oriented magnet. What happens then? Do they hold their breath and refuse to pass through, because they are vertically oriented spin-up electrons and we're forcing them to be measured along a horizontal axis?

No. Instead, the second magnet splits the spin-up electrons into two beams. Half of them are deflected to the right (along the direction of the second magnet) and half of them are deflected to the left.

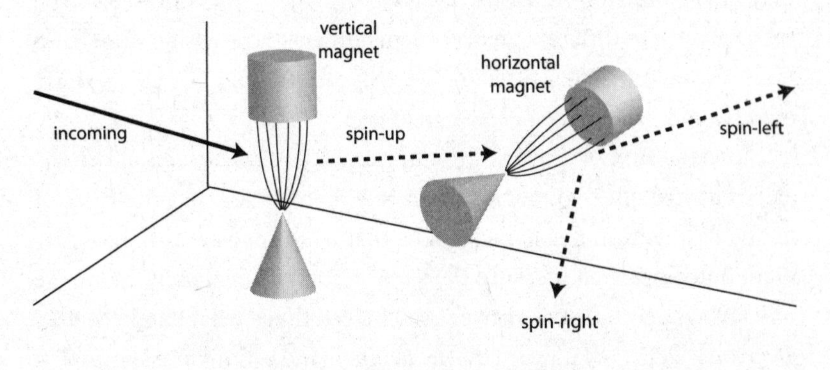

Madness. Our classical intuition makes us think that there is something called "the axis around which the electron is spinning," and it makes sense (maybe) that the spin around *that* axis is quantized. But the experiments show that the axis around which the spin is quantized isn't predetermined by the particle itself; you can choose any axis you like by rotating your magnet appropriately, and the spin will be quantized with respect to that axis.

What we're bumping up against is another manifestation of the uncertainty principle. The lesson we learned was that "position" and "momentum" aren't properties that an electron has; they are just things we can measure about it. In particular, no particle can have a definite value of both simultaneously. Once we specify the exact wave function for position, the probability of observing any particular momentum is entirely fixed, and vice versa.

The same is true for "vertical spin" and "horizontal spin."* These are not separate properties an electron can have; they are just different quantities we can measure. If we express the quantum state in terms of the vertical spin, the probability of observing left or right horizontal spin is entirely fixed. The measurement outcomes we can get are determined by the underlying quantum state, which can be expressed in different but equivalent ways. The uncertainty principle expresses the fact that there are different incompatible measurements we can make on any particular quantum state.

○ ○ ○

Systems with two possible measurement outcomes are so common and useful in quantum mechanics that they are given a cute name: *qubits*. The idea is that a classical "bit" has just two possible values, say, 0 and 1. A qubit (quantum bit) is a system that has two possible measurement outcomes, say, spin-up and spin-down along some specified axis. The state of a generic qubit is a superposition of both possibilities, each weighted by a complex number, the amplitude for each alternative.

---

* And for the third perpendicular direction, which we might call "forward spin," though we didn't measure that.

Quantum computers manipulate qubits in the same way that ordinary computers manipulate classical bits.

We can write the wave function of a qubit as

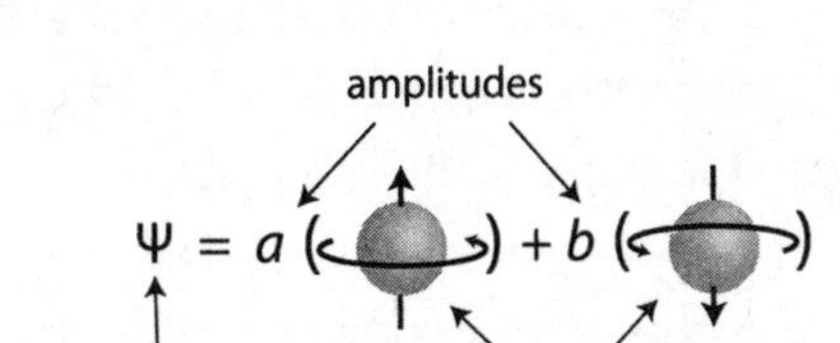

The symbols a and b are complex numbers, representing the amplitudes for spin-up and spin-down, respectively. The pieces of the wave function representing the different possible measurement outcomes, in this case spin-up/-down, are the "components." In this state, the probability of observing the particle to be spin-up would be $|a|^2$, and the probability for spin-down would be $|b|^2$. If, for example, a and b were both equal to the square root of 1/2, the probability of observing spin-up or spin-down would be 1/2.

Qubits can help us understand a crucial feature of wave functions: they are like the hypotenuse of a right triangle, for which the shorter sides are the amplitudes for each possible measurement outcome. In other words, the wave function is like a *vector*—an arrow with a length and a direction.

The vector we're talking about doesn't point in a direction in real physical space, like "up" or "north." Rather, it points in a space defined by all possible measurement outcomes. For a single spin qubit, that's either spin-up or spin-down (once we choose some axis along which to measure). When we say "the qubit is in a superposition of spin-up and spin-down," what we really mean is "the vector representing the quantum state has some component in the spin-up direction, and another component in the spin-down direction."

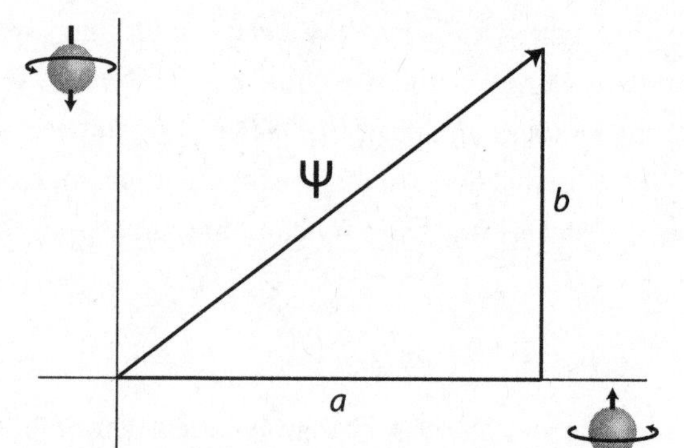

It's natural to think of spin-up and spin-down as pointing in opposite directions. I mean, just look at the arrows. But as quantum states, they are perpendicular to each other: a qubit that is completely spin-up has no component of spin-down, and vice versa. Even the wave function for the position of a particle is a vector, though we normally visualize it as a smooth function throughout space. The trick is to think of every point in space as defining a different component, and the wave function is a superposition of all of them. There are an infinite number of such vectors, so the space of all possible quantum states, called *Hilbert space*, is infinite-dimensional for the position of a single particle. That's why qubits are so much easier to think about. Two dimensions are easier to visualize than infinite dimensions.

When there are only two components in our quantum state, as opposed to infinitely many, it can be hard to think of the state as a "wave function." It's not very wavy, and it doesn't look like a smooth function of space. The right way to think about it is actually the other way around. The quantum state is not a function of ordinary space, it's a function of the abstract "space of measurement outcomes," which for a qubit only includes two possibilities. When the thing we observe is the location of a single particle, the quantum state assigns an amplitude to every possible location, which looks just like a wave in ordinary space. That's the unusual case, however; the wave function is something more abstract, and when more than one particle is involved, it becomes hard to visualize. But we're stuck with the "wave function" terminology. Qubits are great because at least the wave function has only two components.

∘ ∘ ∘

This may seem like an unnecessary mathematical detour, but there are immediate payoffs to thinking about wave functions as vectors. One is explaining the Born rule, which says that the probability for any particular measurement outcome is given by its amplitude squared. We'll dive into details later, but it's easy to see why the idea makes sense. As a vector, the wave function has a length. You might expect that the length could shrink or grow over time, but it doesn't; according to Schrödinger's equation, the wave function just changes its "direction" while maintaining a constant length. And we can compute that length using Pythagoras's theorem from high-school geometry.

The numerical value of the length of the vector is irrelevant; we can just pick it to be a convenient number, knowing that it will remain constant. Let's pick it to be one: every wave function is a vector of length one. The vector itself is just like the hypotenuse of a right triangle, with the components forming the shorter sides. So from Pythagoras's theorem, we have a simple relationship: the squares of the amplitudes add up to unity, $|a|^2 + |b|^2 = 1$.

That's the simple geometric fact underlying the Born rule for quantum probabilities. Amplitudes themselves don't add up to one, but their squares do. That is exactly like an important feature of probability: the sum of probabilities for different outcomes needs to equal one. (Something has to happen, and the total probability of all exclusive somethings adds up to unity.) Another rule is that probabilities need to be non-negative numbers. Once again, amplitudes squared fit the bill:

amplitudes can be negative (or complex), but their squares are non-negative real numbers.

So even before thinking too hard, we can tell that "amplitudes squared" have the right properties to be the probabilities of outcomes—they are a set of non-negative numbers that always add up to one, because that's the length of the wave function. This is at the heart of the whole matter: the Born rule is essentially Pythagoras's theorem, applied to the amplitudes of different branches. That's why it's the amplitudes squared, not the amplitudes themselves or the square root of the amplitudes or anything crazy like that.

The vector picture also explains the uncertainty principle in an elegant way. Remember that spin-up electrons split fifty-fifty into right- and left-spinning electrons when they passed through a subsequent horizontal magnet. That suggests that an electron in a spin-up state is equivalent to a superposition of spin-right and spin-left electron states, and likewise for spin-down.

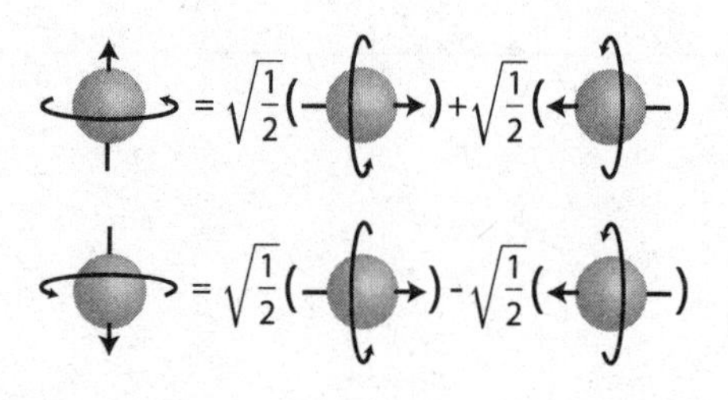

So the idea of being spin-left or spin-right isn't independent from being spin-up or spin-down; any one possibility can be thought of as a superposition of the others. We say that spin-up and spin-down together form a *basis* for the state of a qubit—any quantum state can be written as a superposition of those two possibilities. But spin-left and spin-right form another basis, distinct but equally good. Writing it one way completely fixes the other way.

Think of this in vector terms. If we draw a two-dimensional plane with spin-up as the horizontal axis and spin-down as the vertical axis, from the above relations we see that spin-right and spin-left point at 45 degrees with respect to them. Given any wave function, we could express it in the up/down basis, but we could equally well express it in the right/left basis. One set of axes is rotated with respect to the other, but they are both perfectly legitimate ways of expressing any vector we like.

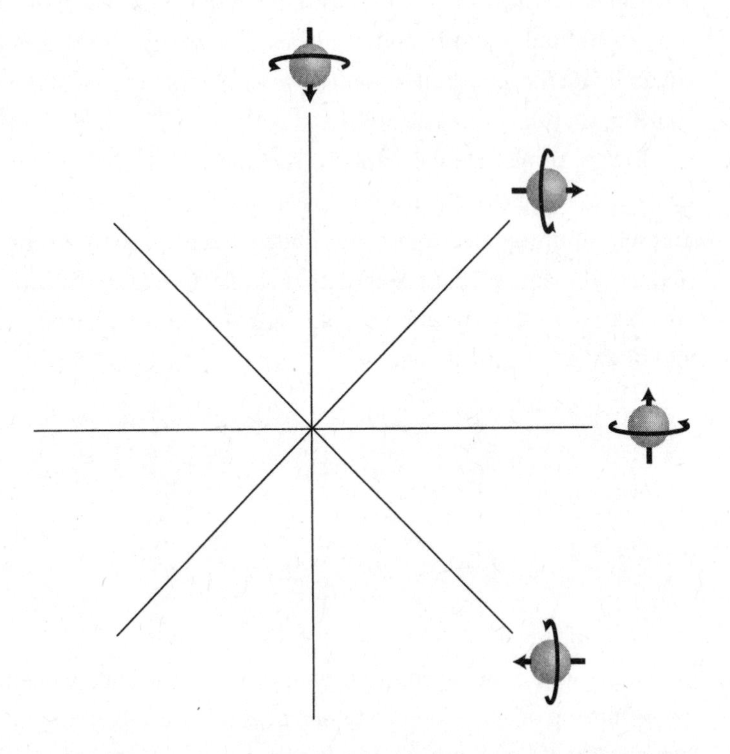

Now we can see where the uncertainty principle comes from. For a single spin, the uncertainty principle says that the state can't have a definite value for the spin along the original axes (up/down) and the rotated axes (right/left) at the same time. This is clear from the picture: if the state is purely spin-up, it's automatically some combination of spin-left and spin-right, and vice versa.

Just as there are no quantum states that are simultaneously localized in position and momentum, there are no states that are simultaneously localized in both vertical spin and horizontal spin. The uncertainty principle reflects the relationship between what really exists (quantum states) and what we can measure (one observable at a time).

Entangled Up in Blue

Wave Functions of Many Parts

Popular discussions of the Einstein-Bohr debates often give the impression that Einstein couldn't quite handle the uncertainty principle, and spent his time trying to invent clever ways to circumvent it. But what really bugged him about quantum mechanics was its apparent nonlocality—what happens at one point in space can seemingly have immediate consequences for experiments done very far away. It took him a while to codify his concerns into a well-formulated objection, and in doing so he helped illuminate one of the most profound features of the quantum world: the phenomenon of *entanglement*.

Entanglement arises because there is only one wave function for the entire universe, not separate wave functions for each piece of it. How do we know that? Why can't we just have a wave function for every particle or field?

Consider an experiment in which we shoot two electrons at each other, moving with equal and opposite velocities. Because both have a negative electric charge, they will repel each other. Classically, if we were given the initial positions and velocities of the electrons, we could calculate precisely the directions into which each of them would scatter. Quantum-mechanically, all we can do is calculate the probability that they will each be observed on various paths after they interact with each other. The wave function of each particle spreads out in a roughly spherical pattern, until we ultimately observe it and pin down a definite direction in which it was moving.

When we actually do this experiment, and observe the electrons after they have scattered, we notice something important. Since the electrons initially had equal and opposite velocities, the total momentum was zero. And momentum is conserved, so the post-interaction momentum should also be zero. This means that while the electrons might emerge moving in various different directions, whatever direction one of them moves in, the other moves in precisely the opposite.

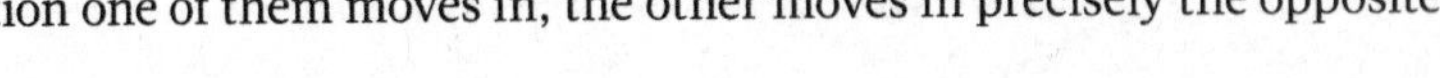

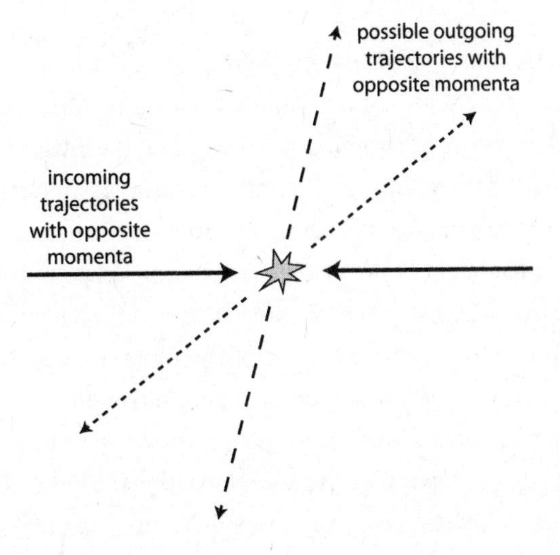

That's funny, when you think about it. The first electron has a probability of scattering at various angles, and so does the second one. But if they each had a separate wave function, those two probabilities would be completely unrelated. We could imagine just observing one of the electrons, and measuring the direction in which it's moving. The other one would be undisturbed. How could it know that it's supposed to be moving in the opposite direction when we actually do measure it?

We've already given away the answer. The two electrons don't have separate wave functions; their behavior is described by the single wave function of the universe. In this case we can ignore the rest of the universe, and just focus in on these two electrons. But we can't ignore one of the electrons and focus in on the other; the predictions we make for observations of either one can be dramatically affected by the outcome of observations of the other. The electrons are entangled.

A wave function is an assignment of a complex number, the amplitude, to each possible observational outcome, and the square of the amplitude equals the probability that we would observe that outcome were we to make that measurement. When we're talking about more than one particle, that means we assign an amplitude to every possible outcome of observing all the particles at once. If what we're observing is positions, for example, the wave function of the universe can be thought of as assigning an amplitude to every possible combination of positions for all the particles in the universe.

You might wonder whether it's possible to visualize something like that. We can do it for the simple case of a single particle that we imagine only moves along one dimension, say, an electron confined to a thin copper wire: we draw a line representing the position of the particle, and plot a function representing the amplitude for each position. (Generally we cheat even in this simple context by just plotting a real number rather than a complex number, but so be it.) For two particles confined to the same one-dimensional motion, we could draw a two-dimensional plane representing the positions of each of the two particles, and then do a three-dimensional contour plot for the wave function. Note that this isn't one particle in two-dimensional space; it's two particles, each on a one-dimensional space, so the wave function is defined on the two-dimensional plane describing both positions.

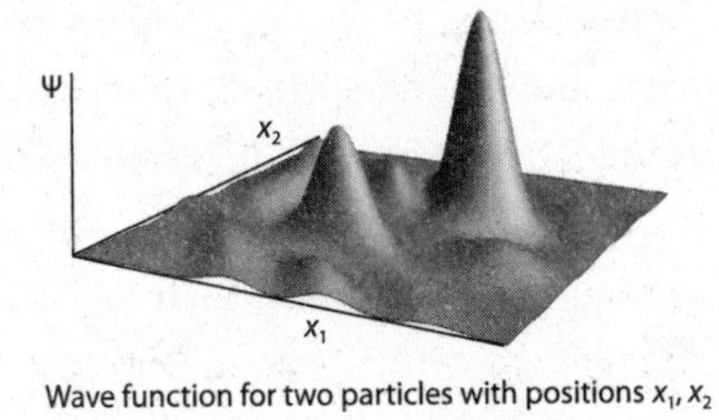

Wave function for one particle with position x Wave function for two particles with positions x_1, x_2

Because of the finite speed of light and a finite time since the Big Bang, we can see only a finite region of the cosmos, which we label "the observable universe." There are approximately 10^{88} particles in the observable universe, mostly photons and neutrinos. That is a number much greater than two. And each particle is located in three-dimensional space, not just a one-dimensional line. How in the world are we supposed to visualize a wave function that assigns an amplitude to every possible configuration of 10^{88} particles distributed through three-dimensional space?

We're not. Sorry. The human imagination wasn't designed to visualize the enormously big mathematical spaces that are routinely used in quantum mechanics. For just one or two particles, we can muddle through; more than that, and we have to describe things in words and equations. Fortunately, the Schrödinger equation is straightforward and definite in what it says about how the wave function behaves. Once we understand what's going on for two particles, the generalization to 10^{88} particles is just math.

∘ ∘ ∘

The fact that wave functions are so big can make thinking about them a little unwieldy. Happily we can cast almost everything interesting to say about entanglement into the much simpler context of just a few qubits.

Borrowing from a whimsical tradition in the literature on cryptography, quantum physicists like to consider two people named Alice and Bob who share qubits with each other. So let's imagine two electrons, *A* belonging to Alice and *B* belonging to Bob. The spins of those two electrons constitute a two-qubit system, and are described by a corresponding wave function. The wave function assigns an amplitude to each configuration of the system as a whole, with respect to something we might observe about it, such as its spin in the vertical direction. So there are four possible measurement outcomes: both spins are up, both spins are down, *A* is up and *B* is down, and *A* is down and *B* is up. The state of the system is some superposition of these four possibilities, which are the basis states. Within each set of parentheses, the first spin is Alice's, and the second is Bob's.

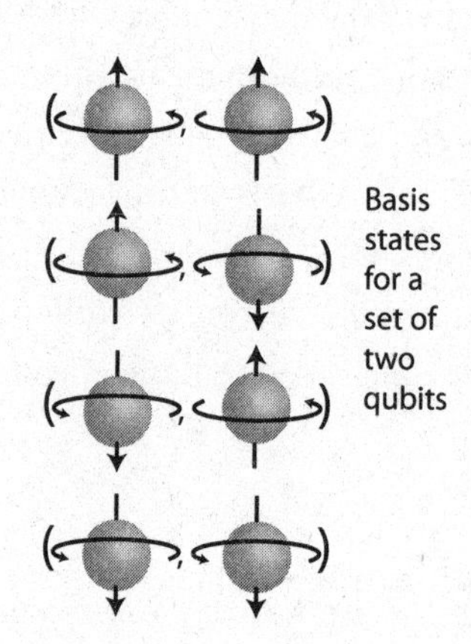

Just because we have two qubits, it doesn't mean they are necessarily entangled. Consider a state that is simply one of the basis states, say, the one where both qubits are spin-up. If Alice measures her qubit along the vertical axis, she will always obtain spin-up, and likewise for Bob. If Alice measures her spin along the horizontal axis, she has a fifty-fifty chance of getting spin-right or spin-left, and again likewise for Bob. But in each case, we don't learn anything about what Bob will see by learning what Alice saw. That's why we can often casually speak of "the wave function of a particle," even though we know better—when different parts of the system are unentangled with each other, it's just as if they have their own wave functions.

Instead, let's consider an equal superposition of two basis states, one with both spins up, and the other with both spins down:

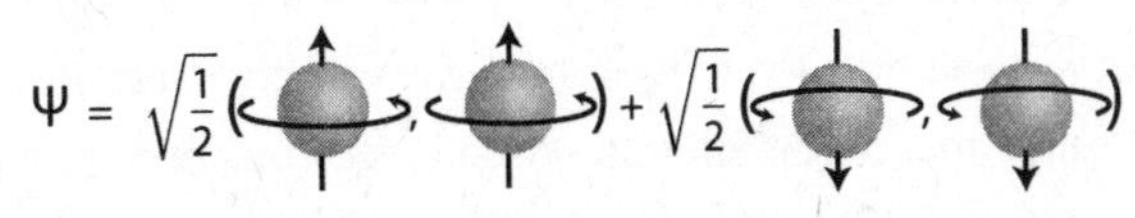

If Alice measures her spin along the vertical axis, she has a fifty-fifty chance of getting spin-up or spin-down, and likewise for Bob. The difference now is that if we learn Alice's outcome before Bob does his measurement, we know what Bob will see with 100 percent confidence—he's going to see the same thing that Alice did. In the language of textbook quantum mechanics, Alice's measurement collapses the wave function onto one of the two basis states, leaving Bob with a deterministic outcome. (In Many-Worlds language, Alice's measurement branches the wave function, creating two different Bobs, each of whom will get a certain outcome.) That's entanglement in action.

◦ ◦ ◦

In the aftermath of the 1927 Solvay Conference, Einstein remained convinced that quantum mechanics, especially as interpreted by the Copenhagen school, did a very good job at making predictions for experimental outcomes, but fell well short as a complete theory of the physical world. His concerns were finally written up for publication in 1935 with his collaborators Boris Podolsky and Nathan Rosen, in a paper that is universally known as simply *EPR*. Einstein later said that the primary ideas had been his, Rosen had done the calculations, and Podolsky had done much of the writing.

EPR considered the position and momentum of two particles moving in opposite directions, but it's easier for us to talk about qubits. Consider two spins that are in the entangled state written above. (It's very easy to create such a state in the lab.) Alice stays home with her qubit, but Bob takes his and embarks on a long journey—say, he jumps in a rocket ship and flies to Alpha Centauri, four light-years away. The entanglement between two particles doesn't fade away as they are moved apart; as long as neither Alice nor Bob measures the spins of their qubits, the overall quantum state will remain the same.

Once Bob arrives safely at Alpha Centauri, Alice finally does measure the spin of her particle, along an agreed-upon vertical axis. Before that measurement, we were completely unsure what such an observation would reveal for her spin, and likewise for Bob's. Let's suppose that Alice observes spin-up. Then, by the rules of quantum mechanics, we immediately know that Bob will also observe spin-up, whenever he gets around to doing a measurement.

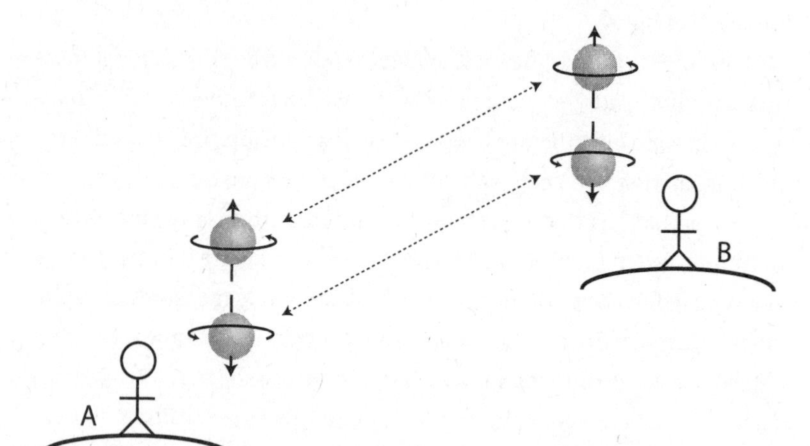

That's weird. Thirty years earlier, Einstein had established the rules of the special theory of relativity, which says among other things that signals cannot travel faster than the speed of light. And yet here we're saying that according to quantum mechanics, a measurement that Alice does here and now has an immediate effect on Bob's qubit, even though it's four light-years away. How does Bob's qubit *know* that Alice's has been measured, and what the outcome was? This is the "spooky action at a distance" that Einstein so memorably fretted about.

It's not necessarily as bad as it seems. The first thing you might wonder about, upon being informed that quantum mechanics apparently sends influences faster than the speed of light, is whether or not we could take advantage of this phenomenon to communicate instantly across large distances. Can we build a quantum-entanglement phone, for which the speed of light is not a limitation at all?

No, we can't. This is pretty clear in our simple example: if Alice measures spin-up, she instantly knows that Bob will also measure spin-up when he gets around to it. But Bob doesn't know that. In order for him to know what the spin of his particle is, Alice has to send him her measurement result by conventional means—which are limited by the speed of light.

You might think there's a loophole: What if Alice doesn't just measure her qubit and find out a random answer, but rather forces her answer to be spin-up? Then Bob would also get spin-up. That would seem like information had been transmitted instantaneously.

The problem is that there's no straightforward way to start with a quantum system that is in a superposition and measure it in such a way that we can force a particular answer. If Alice simply measures her spin, she'll get up or down with equal probabilities, no ifs, ands, or buts. What Alice can do is to manipulate her spin before she measures it, forcing it to be 100 percent spin-up rather than in a superposition. For example, she can shoot a photon at her electron, with just the right properties that the photon leaves the electron alone if the electron was spin-up, and flips the electron to spin-up if it was spin-down. Now Alice's original electron will definitely be measured to be spin-up. But that electron is also no longer entangled with Bob's electron. Rather, the entanglement has been transferred to the photon, which is in a superposition of "left Alice's electron alone" and "bumped into Alice's electron." Bob's electron is completely unaffected, and he's going to get spin-up or spin-down with fifty-fifty probability, so no information has been transmitted.

This is a general feature of quantum entanglement: the *no-signaling theorem*, according to which an entangled pair of particles cannot actually be used to transmit information between two parties faster than light. So quantum mechanics seems to be exploiting a subtle loophole, violating the spirit of relativity (nothing travels faster than the speed of light) while obeying the letter of the law (actual physical particles, and whatever useful information they might convey, cannot travel faster than the speed of light).

◦ ◦ ◦

The so-called *EPR paradox* (which isn't a paradox at all, just a feature of quantum mechanics) goes beyond simple worries about spooky action at a distance. Einstein aimed to show not only that quantum mechanics was spooky but that it couldn't possibly be a complete theory—that there had to be some underlying comprehensive model for which quantum mechanics was simply a useful approximation.

EPR believed in the principle of locality—the physical quantities describing nature are defined at specific points in spacetime, not spread out all over the place, and they interact directly only with other quantities nearby, not at a distance. Said another way, given the speed-of-light restriction of special relativity, locality would seem to imply that nothing we can do to a particle at one location can instantaneously affect measurements we might perform on another particle very far away.

On the face of it, the fact that two widely separated particles can be entangled seems to imply that locality is violated in quantum mechanics. But EPR wanted to be a little more thorough, and establish that there wasn't some clever work-around that would make everything seem local.

They suggested the following principle: if we have a physical system in a specified state, and there is a measurement we can do on that system such that we know with 100 percent certainty what the outcome will be, we associate an *element of reality* with that measurement outcome. In classical mechanics, the position and the momentum of each particle qualify as elements of reality. In quantum mechanics, if we have a qubit in a pure spin-up state, there is an element of reality corresponding to the spin in the vertical direction, but there need not be an element of reality corresponding to the horizontal spin, as we don't know what we will get when we measure that. A "complete" theory, in the EPR formulation, is one in which every element of reality has a direct counterpart in the theory itself, and they argued that quantum mechanics couldn't be complete by this criterion.

Let's take Alice and Bob and their entangled qubits, and imagine that Alice has just measured the vertical spin of her particle, finding that it points upward. We now know that Bob will also measure spin-up, even if Bob doesn't know it himself. So by EPR's lights, there is an element of reality attached to Bob's particle, saying that the spin is up. It's not that this element of reality came into existence when Alice did her measurement, as Bob's particle is very far away, and locality says that the element of reality must be located where the particle is; it must have been there all along.

But now imagine that Alice didn't do the vertical-spin measurement at all, but instead measured the spin of her particle along the horizontal axis. Let's say she measures spin-right for the particle. The entangled quantum state we started with ensures us that Bob will get the same result that Alice did, no matter what direction she chooses to measure her spin in. So we know that Bob would also measure spin-right, and by EPR's lights there is—and was all along—an element of reality that says "spin-right for Bob's qubit if it's measured along the horizontal axis."

There's no way for either Alice's particle or Bob's to know ahead of time which measurement Alice was going to make. Hence, Bob's qubit must come equipped with elements of reality guaranteeing that its spin would be up if measured vertically, and right if measured horizontally.

That's exactly what the uncertainty principle says cannot happen. If the vertical spin is exactly determined, the horizontal spin is completely unknown, and vice versa, at least according to the conventional rules of quantum mechanics. There is nothing in the quantum formalism that can determine both a vertical spin and a horizontal spin at the same time. Therefore, EPR triumphantly conclude, there must be something missing—quantum mechanics cannot be a complete description of physical reality.

The EPR paper caused a stir that reached far beyond the community of professional physicists. The *New York Times*, having been tipped off by Podolsky, published a front-page story about the ideas. This outraged Einstein, who penned a stern letter that the *Times* published, in which he decried advance discussion of scientific results in the "secular press." It's been said that he never spoke to Podolsky again.

EINSTEIN ATTACKS QUANTUM THEORY

Scientist and Two Colleagues Find It Is Not 'Complete' Even Though 'Correct.'

SEE FULLER ONE POSSIBLE

Believe a Whole Description of 'the Physical Reality' Can Be Provided Eventually.

(Courtesy of Wikipedia)

The response from professional scientists was also rapid. Niels Bohr wrote a quick reply to the EPR paper, which many physicists claimed resolved all the puzzles. What is less clear is precisely how Bohr's paper was supposed to have achieved that; as brilliant and creative as he was as a thinker, Bohr was never an especially clear communicator, as he himself admitted. His paper was full of sentences like "in this stage there arises the essential problem of an influence on the precise conditions which define the possible types of prediction which regard the subsequent behavior of the system." Roughly, his argument was that we shouldn't go about attributing elements of reality to systems without taking into account how they are going to be observed. What is real, Bohr seems to suggest, depends not only on what we measure, but on how we choose to measure it.

∘ ∘ ∘

Einstein and his collaborators laid out what they took to be reasonable criteria for a physical theory—locality, and associating elements of reality to deterministically predictable quantities—and showed that quantum mechanics was incompatible with them. But they didn't conclude that quantum mechanics was wrong, just that it was incomplete. The hope remained alive that we would someday find a better theory that both was local and respected reality.

That hope was definitively squashed by John Stewart Bell, a physicist from Northern Ireland who worked at the CERN laboratory in Geneva, Switzerland. He became interested in the foundations of quantum mechanics in the 1960s, at a point in physics history when it was considered thoroughly disreputable to spend time thinking about such things. Today Bell's theorem on entanglement is considered one of the most important results in physics.

The theorem asks us to once again consider Alice and Bob and their entangled qubits with aligned spins. (Such quantum states are now known as *Bell states*, although it was David Bohm who first conceptualized the EPR puzzle in these terms.) Imagine that Alice measures the vertical spin of her particle, and obtains the result that it is spin-up. We now know that if Bob measures the vertical spin of his particle, he will also obtain spin-up. Furthermore, by the ordinary rules of quantum mechanics we know that if Bob chooses to measure the horizontal spin instead, he will get spin-right and spin-left with fifty-fifty probability. We can say that if Bob measures the vertical spin, the correlation between his result and Alice's will be 100 percent (we know exactly what he'll get), whereas if he measures horizontal spin, there will be 0 percent correlation (we have no idea what he will get).

So what if Bob, growing bored all by himself in a spaceship orbiting Alpha Centauri, decides to measure the spin of his particle along some axis in between the horizontal and vertical? (For convenience imagine that Alice and Bob actually share a large number of entangled Bell pairs, so they can keep doing these measurements over and over, and we only care about what happens when Alice observes spin-up.) Then Bob will usually, but not always, observe the spin to be pointed along whatever direction is more closely aligned with the vertical "up." In fact, we can do the math: if Bob's axis is at 45 degrees, exactly halfway between vertical and horizontal, there will be a 71 percent correlation between his results and Alice's. (That's one over the square root of two, if you're wondering where the number comes from.)

What Bell showed, under certain superficially reasonable assumptions, is that this quantum-mechanical prediction is impossible to reproduce in any local theory. In fact, he proved a strict inequality: the best you can possibly do without some kind of spooky action at a distance would be to achieve a 50 percent correlation between Alice and Bob if their measurements were rotated by 45 degrees. The quantum prediction of 71 percent correlation violates Bell's inequality. There is a distinct, undeniable difference between the dream of simple underlying local dynamics, and the real-world predictions of quantum mechanics.

○ ○ ○

I presume you are currently thinking to yourself, "Hey, what do you mean that Bell made superficially reasonable assumptions? Spell them out. I'll decide for myself what I find reasonable and what I don't."

Fair enough. There are two assumptions behind Bell's theorem in particular that one might want to doubt. One is contained in the simple idea that Bob "decides" to measure the spin of his qubit along a certain axis. An element of human choice, or free will, seems to have crept into our theorem about quantum mechanics. That's hardly unique, of course; scientists are always assuming that they can choose to measure whatever they want. But really we think that's just a convenient way of talking, and even those scientists are composed of particles and forces that themselves obey the laws of physics. So we can imagine invoking *superdeterminism*—the idea that the true laws of physics are utterly deterministic (no randomness anywhere), and furthermore that the initial conditions of the universe were laid down at the Big Bang in just precisely such a way that certain "choices" are never going to be made. It's conceivable that one could invent a perfectly local superdeterministic theory that would mimic the predictions of quantum entanglement, simply because the universe was prearranged to make it appear that way. This seems unpalatable to most physicists; if you can delicately arrange your theory to do that, it can basically be arranged to do anything you want, and at that point why are we even doing physics? But some smart people are pursuing the idea.

The other potentially doubtable assumption seems uncontroversial at first glance: that measurements have definite outcomes. When you observe the spin of a particle, you get an actual result, either spin-up or spin-down along whatever axis you are measuring it with respect to. Seems reasonable, doesn't it?

But wait. We actually know about a theory where measurements don't have definite outcomes—austere, Everettian quantum mechanics.

There, it's simply not true that we get either up or down when we measure an electron's spin; in one branch of the wave function we get up, in the other we get down. The universe as a whole doesn't have any single outcome for that measurement; it has multiple ones. That doesn't mean that Bell's theorem is wrong in Many-Worlds; mathematical theorems are unambiguously right, given their assumptions. It just means that the theorem doesn't apply. Bell's result does not imply that we have to include spooky action at a distance in Everettian quantum mechanics, as it does for boring old single-world theories. The correlations don't come about because of any kind of influence being transmitted faster than light, but because of branching of the wave function into different worlds, in which correlated things happen.

For a researcher in the foundations of quantum mechanics, the relevance of Bell's theorem to your work depends on exactly what it is you're trying to do. If you have devoted yourself to the task of inventing a new version of quantum mechanics from scratch, in which measurements do have definite outcomes, Bell's inequality is the most important guidepost you have to keep in mind. If, on the other hand, you're happy with Many-Worlds and are trying to puzzle out how to map the theory onto our observed experience, Bell's result is an automatic consequence of the underlying equations, not an additional constraint you need to worry about moving forward.

One of the fantastic things about Bell's theorem is that it turns the supposed spookiness of quantum entanglement into a straightforwardly experimental question—does nature exhibit intrinsically nonlocal correlations between faraway particles, or not? You'll be happy to hear that experiments have been done, and the predictions of quantum mechanics have been spectacularly verified every time. There is a tradition in popular media of writing articles with breathless headlines like "Quantum Reality Is Even More Bizarre Than Previously Believed!" But when you look into the results they are actually reporting, it's another experiment that confirms exactly what a competent quantum mechanic would have predicted all along using the theory that had been established by 1927, or at least by 1935. We understand quantum mechanics enormously better now than we did back then, but the theory itself hasn't changed.

Which isn't to say that the experiments aren't important or impressive; they are. The problem with testing Bell's predictions, for example, is that you are trying to make sure that the extra correlations predicted by quantum mechanics couldn't have arisen due to some sneaky preexisting classical correlation. How do we know whether some hidden event in the past secretly affected how we chose to measure our spin, or what the measurement outcome was, or both?

Physicists have gone to great lengths to eliminate these possibilities, and a cottage industry has arisen in doing "loophole-free Bell tests." One recent result wanted to eliminate the possibility that an unknown process in the laboratory worked to influence the choice of how to measure the spin. So instead of letting a lab assistant choose the measurement, or even using a random-number generator sitting on a nearby table, the experiment made that choice based on the polarization of photons emitted from stars many light-years away. If there were some nefarious conspiracy to make the world look quantum-mechanical, it had to have been set up hundreds of years ago, when the light left those stars. It's possible, but doesn't seem likely.

It seems that quantum mechanics is right again. So far, quantum mechanics has always been right.

Part Two

SPLITTING

6

Splitting the Universe

Decoherence and Parallel Worlds

The 1935 Einstein-Podolsky-Rosen (EPR) paper on quantum entanglement, and Niels Bohr's response to it, were the last major public salvos in the Bohr-Einstein debates over the foundations of quantum mechanics. Bohr and Einstein had corresponded about quantum theory soon after Bohr proposed his model of quantized electron orbits in 1913, and their dispute came to a head at the 1927 Solvay Conference. In the popular retelling, Einstein would raise some objection to the rapidly coalescing Copenhagen consensus during conversations at the workshop with Bohr, who would spend the evening fretting about it, and then at breakfast Bohr would triumphantly present his rejoinder to the chastened Einstein. We are told that Einstein simply couldn't come to grips with the fact of the uncertainty principle and the notion that God plays dice with the universe.

That's not what happened. Einstein's primary concerns were not with randomness but with realism and locality. His determination to salvage these principles culminated in the EPR paper and their argument that quantum mechanics must be incomplete. But by that time the public-relations battle had been lost, and the Copenhagen approach to quantum mechanics had been adopted by physicists worldwide, who then set about applying quantum mechanics to technical problems in atomic and nuclear physics, as well as the emerging fields of particle physics and quantum field theory. The implications of the EPR paper itself were largely ignored by the community. Wrestling with the confusions at the heart of quantum theory, rather than working on more tangible physics problems, began to be thought of as a somewhat eccentric endeavor. Something that could occupy the time of formerly productive physicists once they reached a certain age and were ready to abandon real work.

In 1933, Einstein left Germany and took a position at the new Institute for Advanced Study in Princeton, New Jersey, where he would remain until his death in 1955. His technical work after 1935 focused largely on classical general relativity and his search for a unified theory of gravitation and electromagnetism, but he never stopped thinking about quantum mechanics. Bohr would occasionally visit Princeton, where he and Einstein would carry on their dialogue.

John Archibald Wheeler joined the physics faculty at Princeton University, down the road from the Institute and Einstein, as an assistant professor in 1934. In later years Wheeler would become known as one of the world's experts in general relativity, popularizing the terms "black hole" and "wormhole," but in his early career he concentrated on quantum problems. He had briefly studied under Bohr in Copenhagen, and in 1939 he and Bohr published a pioneering paper on nuclear fission. Wheeler had great admiration for Einstein, but he venerated Bohr; as he would later put it, "Nothing has done more to convince me that there once existed friends of mankind with the human wisdom of Confucius and Buddha, Jesus and Pericles, Erasmus and Lincoln, than walks and talks under the beech trees of Klampenborg Forest with Niels Bohr."

Wheeler made an impact on physics in a number of ways, one of which was in the mentoring of talented graduate students, including future Nobel laureates such as Richard Feynman and Kip Thorne. One of those students was Hugh Everett III, who would introduce a dramatically new approach to thinking about the foundations of quantum mechanics. We've already sketched his basic idea—the wave function represents reality, it evolves smoothly, and that evolution leads to multiple distinct worlds when a quantum measurement takes place—but now we have the tools to do it right.

∘ ∘ ∘

Everett's proposal, which eventually became his 1957 PhD thesis at Princeton, can be thought of as the purest incarnation of one of Wheeler's favorite principles—that theoretical physics should be "radically conservative." The idea is that a successful physical theory is one that has been tested against experimental data, but only in regimes that experimenters are actually able to reach. One should be conservative, in the sense that we should start with the theories and principles that are already established as successful, rather than arbitrarily introducing new approaches whenever new phenomena are encountered. But one should also be radical, in the sense that the predictions and implications of our theories should be taken seriously in regimes well outside where they have been tested. The phrases "we should start" and "should be taken seriously" are crucial here; of course new theories are warranted when old ones are shown to blatantly contradict the data, and just because a prediction is taken seriously doesn't mean it shouldn't be revised in light of new information. But Wheeler's philosophy was that we should start prudently, with aspects of nature we believe we understand, and then act boldly, extrapolating our best ideas to the ends of the universe.

Part of Everett's inspiration was the search for a theory of quantum gravity, which Wheeler had recently become interested in. The rest of physics—matter, electromagnetism, the nuclear forces—seems to fit

Hugh Everett III
(Courtesy of the Hugh Everett III Archive at the University of California, Irvine, and Mark Everett)

comfortably within the framework of quantum mechanics. But gravity was (and remains) a stubborn exception. In 1915, Einstein proposed the general theory of relativity, according to which spacetime itself is a dynamical entity whose bends and warps are what you and I perceive as the force of gravity. But general relativity is a thoroughly classical theory, with analogues of position and momentum for the curvature of spacetime, and no limits on how we might measure them. Taking that theory and "quantizing" it, constructing a theory of wave functions of spacetime rather than particular classical spacetimes has proven difficult.

The difficulties of quantum gravity are both technical—calculations tend to blow up and give infinitely big answers—and also conceptual. Even in quantum mechanics, while you might not be able to say precisely where a certain particle is, the notion of "a point in space" is perfectly well defined. We can specify a location and ask what is the probability of finding the particle nearby. But if reality doesn't consist of stuff distributed through space, but rather is a quantum wave function describing superpositions of different possible spacetimes, how do we even ask "where" a certain particle is observed?

The puzzles become worse when we turn to the measurement problem. By the 1950s the Copenhagen school was established doctrine, and physicists had made their peace with the idea of wave functions collapsing when a measurement occurred. They were even willing to go along with treating the measurement process as a fundamental part of our best description of nature. Or, at least, not to fret too much about it.

But what happens when the quantum system under consideration is the entire universe? Crucial to the Copenhagen approach is the distinction between the quantum system being measured and the classical observer doing the measuring. If the system is the universe as a whole, we are all inside it; there's no external observer to whom we can appeal. Years later, Stephen Hawking and others would study quantum cosmology to discuss how a self-contained universe could have an earliest moment in time, presumably identified with the Big Bang.

While Wheeler and others thought about the technical challenges of quantum gravity, Everett became fascinated by these conceptual problems, especially how to handle measurement. The seeds of the Many-Worlds formulation can be traced to a late-night discussion in 1954 with fellow young physicists Charles Misner (also a student of Wheeler's) and Aage Petersen (an assistant of Bohr's, visiting from Copenhagen). All parties agree that copious amounts of sherry were consumed on the occasion.

Clearly, Everett reasoned, if we're going to talk about the universe in quantum terms, we can't carve out a separate classical realm. Every part of the universe will have to be treated according to the rules of quantum mechanics, including the observers within it. There will only be a single quantum state, described by what Everett called the "universal wave function" (and we've been calling "the wave function of the universe").

If everything is quantum, and the universe is described by a single wave function, how is measurement supposed to occur? It must be, Everett reasoned, when one part of the universe interacts with another part of the universe in some appropriate way. That is something that's going to happen automatically, he noticed, simply due to the evolution of the universal wave function according to the Schrödinger equation.

We don't need to invoke any special rules for measurement at all; things bump into each other all the time.

It's for this reason that Everett titled his eventual paper on the subject "'Relative State' Formulation of Quantum Mechanics." As a measurement apparatus interacts with a quantum system, the two become entangled with each other. There are no wave-function collapses or classical realms. The apparatus itself evolves into a superposition, entangled with the state of the thing being observed. The apparently definite measurement outcome ("the electron is spin-up") is only relative to a particular state of the apparatus ("I measured the electron to be spin-up"). The other possible measurement outcomes still exist and are perfectly real, just as separate worlds. All we have to do is to courageously face up to what quantum mechanics has been trying to tell us all along.

○ ○ ○

Let's be a little more explicit about what happens when a measurement is made, according to Everett's theory.

Imagine that we have a spinning electron, which could be observed to be in states of either spin-up or spin-down with respect to some chosen axis. Before measurement, the electron will typically be in some superposition of up and down. We also have a measuring apparatus, which is a quantum system in its own right. Imagine that it can be in superpositions of three different possibilities: it can have measured the spin to be up, it can have measured the spin to be down, or it might not yet have measured the spin at all, which we call the "ready" state.

The fact that the measurement apparatus does its job tells us how the quantum state of the combined spin+apparatus system evolves according to the Schrödinger equation. Namely, if we start with the apparatus in its ready state and the spin in a purely spin-up state, we are guaranteed that the apparatus evolves to a pure measured-up state, like so:

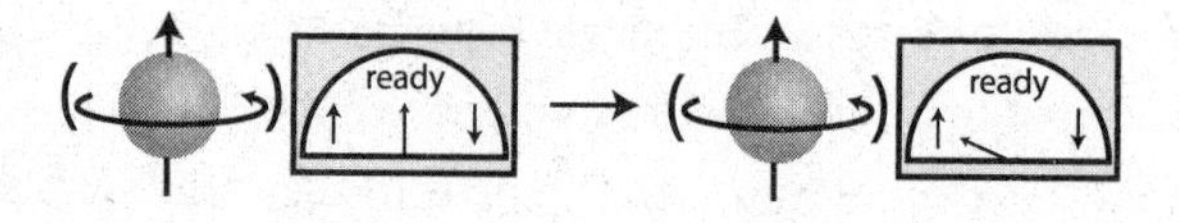

The initial state on the left can be read as "the spin is in the up state, and the apparatus is in its ready state," while the one on the right, where the pointer indicates the up arrow, is "the spin is in the up state, and the apparatus has measured it to be up."

Likewise, the ability to successfully measure a pure-down spin implies that the apparatus must evolve from "ready" to "measured down":

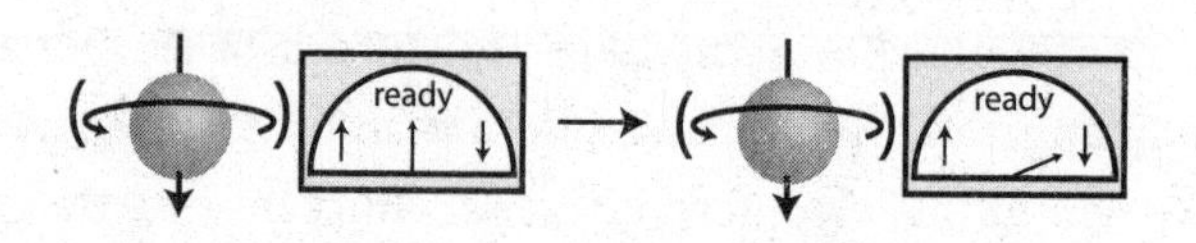

What we want, of course, is to understand what happens when the initial spin is not in a pure up or down state, but in some superposition of both. The good news is that we already know everything we need. The rules of quantum mechanics are clear: if you know how the system evolves starting from two different states, the evolution of a superposition of both those states will just be a superposition of the two evolutions. In other words, starting from a spin in some superposition and the measurement device in its ready state, we have:

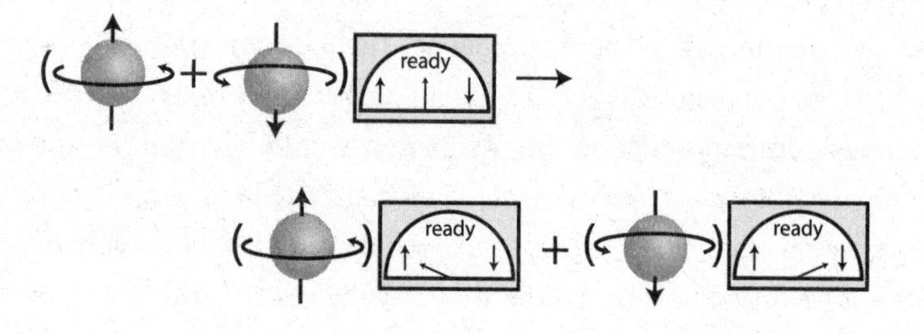

The final state now is an entangled superposition: the spin is up and it was measured to be up, plus the spin is down and it was measured to be down. At this point it's not strictly correct to say "the spin is in a superposition" or "the apparatus is in a superposition." Entanglement prevents us from talking about the wave function of the spin, or that of the apparatus, individually, because what we will observe about one can depend on what we observe about the other. The only thing we can say is "the spin+apparatus system is in a superposition."

This final state is the clear, unambiguous, definitive final wave function for the combined spin+apparatus system, if all we do is evolve it according to the Schrödinger equation. This is the secret to Everettian quantum mechanics. The Schrödinger equation says that an accurate measuring apparatus will evolve into a macroscopic superposition, which we will ultimately interpret as branching into separate worlds. We didn't put the worlds in; they were always there, and the Schrödinger equation inevitably brings them to life. The problem is that we never seem to come across superpositions involving big macroscopic objects in our experience of the world.

The traditional remedy has been to monkey with the fundamental rules of quantum mechanics in one way or another. Some approaches say that the Schrödinger equation isn't always applicable, others say that there are additional variables over and above the wave function. The Copenhagen approach is to disallow the treatment of the measurement apparatus as a quantum system in the first place, and treat wave function collapse as a separate way the quantum state can evolve. One way or another, all of these approaches invoke contortions in order to not accept superpositions like the one written above as the true and complete description of nature. As Everett would later put it, "The Copenhagen Interpretation is hopelessly incomplete because of its *a priori* reliance on classical physics . . . as well as a philosophic monstrosity with a 'reality' concept for the macroscopic world and denial of the same for the microcosm."

Everett's prescription was simple: stop contorting yourself. Accept the reality of what the Schrödinger equation predicts. Both parts of the final wave function are actually there. They simply describe separate, never-to-interact-again worlds.

Everett didn't introduce anything new into quantum mechanics; he removed some extraneous clunky pieces from the formalism. Every non-Everettian version of quantum mechanics is, as physicist Ted Bunn has put it, a "disappearing worlds" theory. If the multiple worlds bother you, you have to fiddle with either the nature of quantum states or their ordinary evolution in order to get rid of them. Is it worth it?

° ° °

There's a looming question here. We're familiar with how wave functions represent superpositions of different possible measurement outcomes. The wave function of an electron can put it in a superposition of various possible locations, as well as in a superposition of spin-up and spin-down. But we were never tempted to say that each part of the superposition was a separate "world." Indeed, it would have been incoherent to do so. An electron that is in a pure spin-up state with respect to the vertical axis is in a superposition of spin-up and spin-down with respect to the horizontal axis. So does that describe one world, or two?

Everett suggested that it is logically consistent to think of superpositions involving macroscopic objects as describing separate worlds. But at the time he was writing, physicists hadn't yet developed the technical tools necessary to turn this into a complete picture. That understanding only came later, with the appreciation of a phenomenon known as *decoherence*. Introduced in 1970 by the German physicist Hans Dieter Zeh, the idea of decoherence has become a central part of how physicists think about quantum dynamics. To the modern Everettian, decoherence is absolutely crucial to making sense of quantum mechanics. It explains once and for all why wave functions seem to collapse when you measure quantum systems—and indeed what a "measurement" really is.

We know there is only one wave function, the wave function of the universe. But when we're talking about individual microscopic particles, they can settle into quantum states where they are unentangled from the rest of the world. In that case, we can sensibly talk about "the wave function of this particular electron" and so forth, keeping in mind that it's really just a useful shortcut we can employ when systems are unentangled with anything else.

With macroscopic objects, things aren't that simple. Consider our spin-measuring apparatus, and let's imagine we put it in a superposition of having measured spin-up and spin-down. The dial of the apparatus includes a pointer that is pointing either to Up or to Down. An apparatus like that doesn't stay separate from the rest of the world. Even if it looks like it's just sitting there, in reality the air molecules in the room are constantly bumping into it, photons of light are bouncing off of it, and so on. Call all that other stuff—the entire rest of the universe—the *environment*. In ordinary situations, there's no way to stop a macroscopic object from interacting with its environment, even if very gently. Such interactions will cause the apparatus to become entangled with the environment, for example, because a photon would reflect off the dial if the pointer is in one position, but be absorbed by it if the pointer is pointing somewhere else.

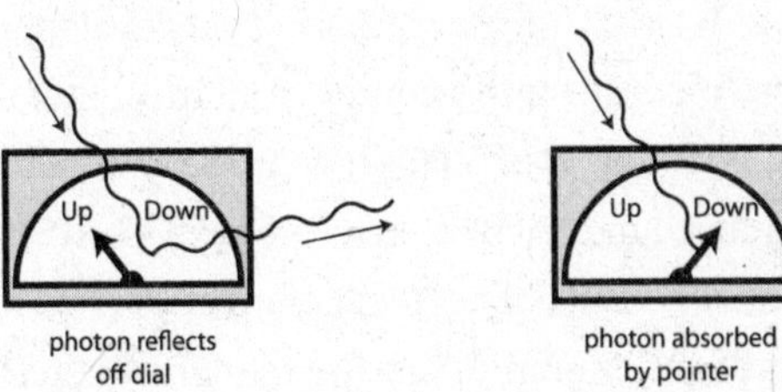

So the wave function we wrote down above, where an apparatus became entangled with a qubit, wasn't quite the whole story. Putting the environment states in curly braces, we should have written

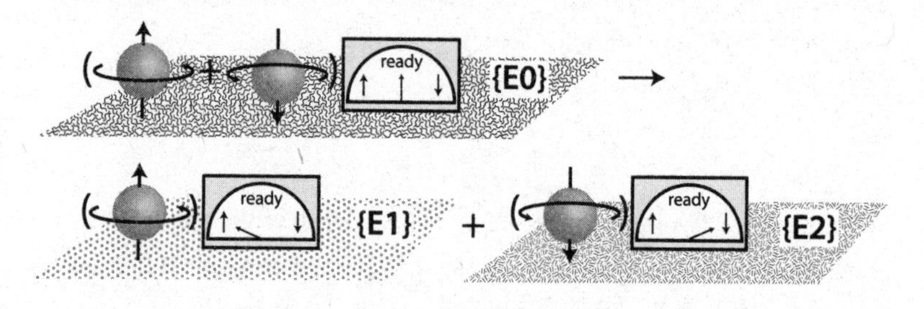

It doesn't really matter what the environment states actually are, so we've portrayed them as different backgrounds labeled **{E0}**, **{E1}**, and **{E2}**. We don't (and generally can't) keep track of exactly what's going on in the environment—it's too complicated. It's not going to just be a single photon that interacts differently with different parts of the apparatus's wave function, it will be a huge number of them. Nobody can be expected to keep track of every photon or particle in a room.

That simple process—macroscopic objects become entangled with the environment, which we cannot keep track of—is decoherence, and it comes with universe-altering consequences. Decoherence causes the wave function to split, or *branch*, into multiple worlds. Any observer branches into multiple copies along with the rest of the universe. After branching, each copy of the original observer finds themselves in a world with some particular measurement outcome. To them, the wave function *seems* to have collapsed. We know better; the collapse is only apparent, due to decoherence splitting the wave function.

We don't know how often branching happens, or even whether that's a sensible question to ask. It depends on whether there are a finite or infinite number of degrees of freedom in the universe, which is currently an unanswered question in fundamental physics. But we do know that there's a lot of branching going on; it happens every time a quantum system in a superposition becomes entangled with the environment. In a typical human body, about 5,000 atoms undergo radioactive decay every second. If every decay branches the wave function in two, that's 2^{5000} new branches every second. It's a lot.

◦ ◦ ◦

What makes a "world," anyway? We just wrote down a single quantum state describing a spin, an apparatus, and an environment. What makes us say that it describes two worlds, rather than just one?

One thing you would like to have in a world is that different parts of it can, at least in principle, affect each other. Consider the following "ghost world" scenario (not meant as a true description of reality, just a colorful analogy): when living beings die, they all become ghosts. These ghosts can see and talk to one another, but they cannot see or talk to us, nor can we see or talk to them. They live on a separate Ghost Earth, where they can build ghost houses and go to their ghost jobs. But neither they nor their surroundings can interact with us and the stuff around us in any way. In this case it makes sense to say that the ghosts inhabit a truly separate ghost world, for the fundamental reason that what happens in the ghost world has absolutely no bearing on what happens in our world.

Now apply this criterion to quantum mechanics. We're not interested in whether the spin and its measuring apparatus can influence each other—they obviously can. What we care about is whether one component of, say, the apparatus wave function (for example, the piece where the dial is pointing to Up) can possibly influence another piece (for example, where it's pointing to Down). We've previously come across a situation just like this, where the wave function influences itself—in the phenomenon of interference from the double-slit experiment. When we passed electrons through two slits without measuring which one they went through, we saw interference bands on the final screen, and attributed them to the cancellation between the contribution to the total probability from each of the two slits. Crucially, we implicitly assumed that the electron didn't interact and become entangled with anything along its journey; it didn't decohere.

When instead we did detect which slit the electron went through, the interference bands went away. At the time we attributed this to the fact that a measurement had been performed, collapsing the electron's wave function at one slit or another. Everett gives us a much more compelling story to tell.

What actually happened was that the electron became entangled with the detector as it moved through the slits, and then the detector quickly became entangled with the environment. The process is precisely analogous to what happened to our spin above, except that we're measuring whether the electron went through the left slit L or the right slit R:

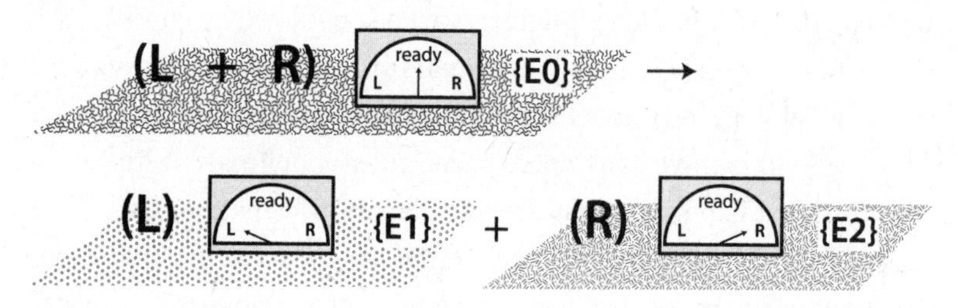

No mysterious collapsing; the whole wave function is still there, evolving cheerfully according to the Schrödinger equation, leaving us in a superposition of two entangled pieces. But note what happens as the electron continues on toward the screen. As before, the state of the electron at any given point on the screen will receive a contribution from what passed through slit L, and another contribution from what passed through slit R. But now those contributions *won't interfere with each other.* In order to get interference, we need to be adding up two equal and opposite quantities:

$$1 + (-1) = 0.$$

But there is no point on the screen where we will find equal and opposite contributions to the electron's wave function from the L and R slits, because passing through those slits entangled the electron with *different states of the rest of the world.* When we say equal and opposite, we mean precisely equal and opposite, not "equal and opposite except for that thing we're entangled with." Being entangled with different states of the detector and environment—being decohered, in other words—means that the two parts of the electron's wave function can no longer interfere with each other. And that means they can't interact at all. And that means they are, for all intents and purposes, part of separate worlds.* From the point of view of things entangled with one branch of the wave function, the other branches might as well be populated by ghosts.

The Many-Worlds formulation of quantum mechanics removes once and for all any mystery about the measurement process and collapse of the wave function. We don't need special rules about making an observation: all that happens is that the wave function keeps chugging along in accordance with the Schrödinger equation. And there's nothing special about what constitutes "a measurement" or "an observer"—a measurement is any interaction that causes a quantum system to become entangled with the environment, creating decoherence and a branching into separate worlds, and an observer is any system that brings such an interaction about. Consciousness, in particular, has nothing to do with

---

* The set of all branches of the wave function is different from what cosmologists often call "the multiverse." The cosmological multiverse is really just a collection of regions of space, generally far away from one another, where local conditions look very different.

it. The "observer" could be an earthworm, a microscope, or a rock. There's not even anything special about macroscopic systems, other than the fact that they can't help but interact and become entangled with the environment. The price we pay for such powerful and simple unification of quantum dynamics is a large number of separate worlds.

◦ ◦ ◦

Everett himself wasn't familiar with decoherence, so his picture wasn't quite as robust and complete as the one we've painted. But his way of rethinking the measurement problem and offering a unified picture of quantum dynamics was compelling from the start. Even in theoretical physics, people do sometimes get lucky, hitting upon an important idea more because they were in the right place at the right time than because they were particularly brilliant. That's not the case with Hugh Everett; those who knew him testify uniformly to his incredible intellectual gifts, and it's clear from his writings that he had a thorough understanding of the implications of his ideas. Were he still alive, he would be perfectly at home in modern discussions of the foundations of quantum mechanics.

What was hard was getting others to appreciate those ideas, and that included his advisor. Wheeler was personally very supportive of Everett, but he was also devoted to his own mentor, Bohr, and was convinced of the basic soundness of the Copenhagen approach. He simultaneously wanted Everett's ideas to get a wide hearing, and to ensure that they weren't interpreted as a direct assault on Bohr's way of thinking about quantum mechanics.

Yet Everett's theory *was* a direct assault on Bohr's picture. Everett himself knew it, and enjoyed illustrating the nature of this assault in vivid language. In an early draft of his thesis, Everett used the analogy of an amoeba dividing to illustrate the branching of the wave function: "One can imagine an intelligent amoeba with a good memory. As time progresses the amoeba is constantly splitting, each time the resulting amoebas having the same memories as the parent. Our amoeba hence does not have a life line, but a life tree." Wheeler was put off by the blatantness of this (quite accurate) metaphor, scribbling in the margin of the manuscript, "Split? Better words needed." Advisor and student were constantly tussling over the best way to express the new theory, with Wheeler advocating caution and prudence while Everett favored bold clarity.

In 1956, as Everett was working on finishing his dissertation, Wheeler visited Copenhagen and presented the new scenario to Bohr and his colleagues, including Aage Petersen. He attempted to present it anyway; by this time the wave-functions-collapse-and-don't-ask-embarrassing-questions-about-exactly-how school of quantum theory had hardened into conventional wisdom, and those who accepted it weren't interested in revisiting the foundations when there was so much interesting applied work to be done. Letters from Wheeler, Everett, and Petersen flew back and forth across the Atlantic, continuing when Wheeler returned to Princeton and helped Everett craft the final form of his dissertation. The agony of this process is reflected in the evolution of the paper itself: Everett's first draft was titled "Quantum Mechanics by the Method of the Universal Wave Function," and a revised version was called "Wave Mechanics Without Probability." This document, later dubbed the "long version" of the thesis, wasn't published until 1973. A "short version" was finally submitted for Everett's PhD as "On the Foundations of Quantum Mechanics," and eventually published in 1957 as "'Relative State' Formulation of Quantum Mechanics." It omitted many of the juicier sections Everett had originally composed, including examinations of the foundations of probability and information theory and an overview of the quantum measurement problem, focusing instead on applications to quantum cosmology. (No amoebas appear in the published paper, but Everett did manage to insert the word "splitting" in a footnote added in proof while Wheeler wasn't looking.)

Furthermore, Wheeler wrote an "assessment" article that was published alongside Everett's, which suggested that the new theory was radical and important, while at the same time attempting to paper over its manifest differences with the Copenhagen approach.

The arguments continued, without much headway being made. It is worth quoting from a letter Everett wrote to Petersen, in which his frustration comes through:

> Lest the discussion of my paper die completely, let me add some fuel to the fire with . . . criticisms of the 'Copenhagen interpretation.' . . . I do not think you can dismiss my viewpoint as simply a misunderstanding of Bohr's position. . . . I believe that basing quantum mechanics upon classical physics was a necessary provisional step, but that the time has come . . . to treat [quantum mechanics] in its own right as a fundamental theory without any dependence on classical physics, and to derive classical physics from it. . . .
>
> Let me mention a few more irritating features of the Copenhagen Interpretation. You talk of the massiveness of macro systems allowing one to neglect further quantum effects (in discussions of breaking the measuring chain), but never give any justification for this flatly asserted dogma. [And] there is nowhere to be found any consistent explanation for this 'irreversibility' of the measuring process. It is again certainly not implied by wave mechanics, nor classical mechanics either. Another independent postulate?

But Everett decided not to continue the academic fight. Before finishing his PhD, he accepted a job at the Weapons Systems Evaluation Group for the US Department of Defense, where he studied the effects of nuclear weapons. He would go on to do research on strategy, game theory, and optimization, and played a role in starting several new companies. It's unclear the extent to which Everett's conscious decision to not apply for professorial positions was motivated by criticism of his upstart new theory, or simply by impatience with academia in general.

He did, however, maintain an interest in quantum mechanics, even if he never published on it again. After Everett defended his PhD and was already working for the Pentagon, Wheeler persuaded him to visit Copenhagen for himself and talk to Bohr and others. The visit didn't go well; afterward Everett judged that it had been "doomed from the beginning."

Bryce DeWitt, an American physicist who had edited the journal where Everett's thesis appeared, wrote a letter to him complaining that the real world obviously didn't "branch," since we never experience such things. Everett replied with a reference to Copernicus's similarly daring idea that the Earth moves around the sun, rather than vice versa: "I can't resist asking: Do you feel the motion of the earth?" DeWitt had to admit that was a pretty good response. After mulling the matter over for a while, by 1970 DeWitt had become an enthusiastic Everettian. He put a great deal of effort into pushing the theory, which had languished in obscurity, toward greater public recognition. His strategies included an influential 1970 article in *Physics Today*, followed by a 1973 essay collection that included at last the long version of Everett's dissertation, as well as a number of commentaries. The collection was called simply *The Many-Worlds Interpretation of Quantum Mechanics*, a vivid name that has stuck ever since.

In 1976, John Wheeler retired from Princeton and took up a position at the University of Texas, where DeWitt was also on the faculty. Together they organized a workshop in 1977 on the Many-Worlds theory, and Wheeler coaxed Everett into taking time off from his defense work in order to attend. The conference was a success, and Everett made a significant impression on the assembled physicists in the audience. One of them was the young researcher David Deutsch, who would go on to become a major proponent of Many-Worlds, as well as an early pioneer of quantum computing. Wheeler went so far as to propose a new research institute in Santa Barbara, where Everett could return to full-time work on quantum mechanics, but ultimately nothing came of it.

Everett died in 1982, age fifty-one, of a sudden heart attack. He had not lived a healthy lifestyle, overindulging in eating, smoking, and drinking. His son, Mark Everett (who would go on to form the band Eels), has said that he was originally upset with his father for not taking better care of himself. He later changed his mind: "I realize that there is a certain value in my father's way of life. He ate, smoked and drank as he pleased, and one day he just suddenly and quickly died. Given some of the other choices I'd witnessed, it turns out that enjoying yourself and then dying quickly is not such a hard way to go."

Order and Randomness

Where Probability Comes From

One sunny day in Cambridge, England, Elizabeth Anscombe ran into her teacher, Ludwig Wittgenstein. "Why do people say," Wittgenstein opened in his inimitable fashion, "that it was natural to think that the sun went round the earth, rather than that the earth turned on its axis?" Anscombe gave the obvious answer, that it just *looks like* the sun goes around the Earth. "Well," Wittgenstein replied, "what would it have *looked like* if the Earth had turned on its axis?"

This anecdote—recounted by Anscombe herself, and which Tom Stoppard retold in his play *Jumpers*—is a favorite among Everettians. Physicist Sidney Coleman used to relate it in lectures, and philosopher of physics David Wallace used it to open his book *The Emergent Multiverse*. It even bears a family resemblance to Hugh Everett's remark to Bryce DeWitt.

It's easy to see why the observation is so relevant. Any reasonable person, when first told about the Many-Worlds picture, has an immediate, visceral objection: it just doesn't *feel like* I personally split into multiple people whenever a quantum measurement is performed. And it certainly doesn't *look like* there are all sorts of other universes existing parallel to the one I find myself in.

Well, the Everettian replies, channeling Wittgenstein: What would it feel and look like if Many-Worlds were true?

The hope is that people living in an Everettian universe would experience just what people actually do experience: a physical world that seems to obey the rules of textbook quantum mechanics to a high degree of accuracy, and in many situations is well approximated by classical mechanics. But the conceptual distance between "a smoothly evolving wave function" and the experimental data it is meant to explain is quite large. It's not obvious that the answer we can give to Wittgenstein's question is the one we want. Everett's theory might be austere in its formulation, but there's still a good amount of work to be done to fully flesh out its implications.

In this chapter we'll confront a major puzzle for Many-Worlds: the origin and nature of probability. The Schrödinger equation is perfectly deterministic. Why do probabilities enter at all, and why do they obey the Born rule: probabilities equal amplitudes—the complex numbers the wave function associates with each possible outcome—squared? Does it even make sense to speak of the probability of ending up on some particular branch if there will be a future version of myself on every branch?

In the textbook or Copenhagen versions of quantum mechanics, there's no need to "derive" the Born rule for probabilities. We just plop it down there as one of the postulates of the theory. Why couldn't we do the same thing in Many-Worlds?

The answer is that even though the rule would sound the same in both cases—"probabilities are given by the wave function squared"—their meanings are very different. The textbook version of the Born rule really is a statement about how often things happen, or how often they will happen in the future. Many-Worlds has no room for such an extra postulate; we know exactly what will happen, just from the basic rule that the wave function always obeys the Schrödinger equation. Probability in

Many-Worlds is necessarily a statement about what we should *believe* and how we should *act*, not about how often things happen. And "what we should believe" isn't something that really has a place in the postulates of a physical theory; it should be implied by them.

Moreover, as we will see, there is neither any room for an extra postulate, nor any need for one. Given the basic structure of quantum mechanics, the Born rule is natural and automatic. Since we tend to see Born rule–like behavior in nature, this should give us confidence that we're on the right track. A framework in which an important result can be derived from more fundamental postulates should, all else being equal, be preferred to one where it needs to be separately assumed.

If we successfully address this question, we will have made significant headway toward showing the world we would expect to see if Many-Worlds were true is the world we actually do see. That is, a world that is closely approximated by classical physics, except for quantum measurement events, during which the probability of obtaining any particular outcome is given by the Born rule.

◦ ◦ ◦

The issue of probabilities is often phrased as trying to derive why probabilities are given by amplitudes squared. But that's not really the hard part. Squaring amplitudes in order to get probabilities is a very natural thing to do; there weren't any worries that it might have been the wave function to the fifth power or anything like that. We learned that back in Chapter Five, when we used qubits to explain that the wave function can be thought of as a vector. That vector is like the hypotenuse of a right triangle, and the individual amplitudes are like the shorter sides of that triangle. The length of the vectors equals one, and by Pythagoras's theorem that's the sum of the squares of all the amplitudes. So "amplitudes squared" naturally look like probabilities: they're positive numbers that add up to one.

The deeper issue is why there is anything unpredictable about Everettian quantum mechanics at all, and if so, why there is any specific rule for attaching probabilities. In Many-Worlds, if you know the wave function at one moment in time, you can figure out precisely what it's going to be at any other time, just by solving the Schrödinger equation. There's nothing chancy about it. So how in the world is such a picture supposed to recover the reality of our observations, where the decay of a nucleus or the measurement of a spin seems irreducibly random?

Consider our favorite example of measuring the spin of an electron. Let's say we start the electron in an equal superposition of spin-up and spin-down with respect to the vertical axis, and send it through a Stern-Gerlach magnet. Textbook quantum mechanics says that we have a 50 percent chance of the wave function collapsing to spin-up, and a 50 percent chance of it collapsing to spin-down. Many-Worlds, on the other hand, says there is a 100 percent chance of the wave function of the universe evolving from one world into two. True, in one of those worlds the experimenter will have seen spin-up and in the other they will have seen spin-down. But both worlds are indisputably *there*. If the question we're asking is "What is the chance I will end up being the experimenter on the spin-up branch of the wave function?," there doesn't seem to be any answer. You will not be one or other experimenters; your current single self will evolve, with certainty, into both of them. How are we supposed to talk about probabilities in such a situation?

It's a good question. To answer it, we have get a bit philosophical, and think about what "probability" really means.

◦ ◦ ◦

You will not be surprised to learn that there are competing schools of thought on the issue of probability. Consider tossing a fair coin. "Fair" means that the coin will come up heads 50 percent of the time and tails

50 percent of the time. At least in the long run; nobody is surprised when you toss a coin twice and it comes up tails both times.

This "in the long run" caveat suggests a strategy for what we might mean by probability. For just a few coin tosses, we wouldn't be surprised at almost any outcome. But as we do more and more, we expect the total proportion of heads to come closer to 50 percent. So perhaps we can define the probability of getting heads as the fraction of times we actually would get heads, if the coin were tossed an infinite number of times.

This notion of what we mean by probability is sometimes called *frequentism*, as it defines probability as the relative frequency of an occurrence in a very large number of trials. It matches pretty well with our intuitive notions of how probability functions when we toss coins, roll dice, or play cards. To a frequentist, probability is an objective notion, since it only depends on features of the coin (or whatever other system we're talking about), not on us or our state of knowledge.

Frequentism fits comfortably with the textbook picture of quantum mechanics and the Born rule. Maybe you don't actually send an infinite number of electrons through a magnetic field to measure their spins, but you could send a very large number. (The Stern-Gerlach experiment is a favorite one to reproduce in undergraduate lab courses for physics majors, so over the years quite a number of spins have been measured this way.) We can gather enough statistics to convince ourselves that the probability in quantum mechanics really is just the wave function squared.

Many-Worlds is a different story. Say we put an electron into an equal superposition of spin-up and spin-down, measure its spin, then repeat a large number of times. At every measurement, the wave function branches into a world with a spin-up result and one with a spin-down. Imagine that we record our results, labeling spin-up as "0" and spin-down as "1." After fifty measurements, there will be a world where the record looks like

10101011111011001011001010100011101100011101000001.

That seems random enough, and to obey the proper statistics: there are twenty-four 0's, and twenty-six 1's. Not exactly fifty-fifty, but as close as we should expect.

But there will also be a world where every measurement returned spin-up, so that the record was just a list of fifty 0's. And a world where all the spins were observed to be down, so the record was a list of fifty 1's. And every other possible string of 0's and 1's. If Everett is right, there is a 100 percent probability that each possibility is realized in some particular world.

In fact, I'll make a confession: there really are such worlds. The random-looking string above wasn't something I made up to look random, nor was it created by a classical random-number generator. It was actually created by a *quantum* random-number generator: a gizmo that makes quantum measurements and uses them to generate random sequences of 0's and 1's. According to Many-Worlds, when I generated that random number, the universe split into 2^{50} copies (that's 1,125,899,906,842,624, or approximately 1 quadrillion), each of which carries a slightly different number.

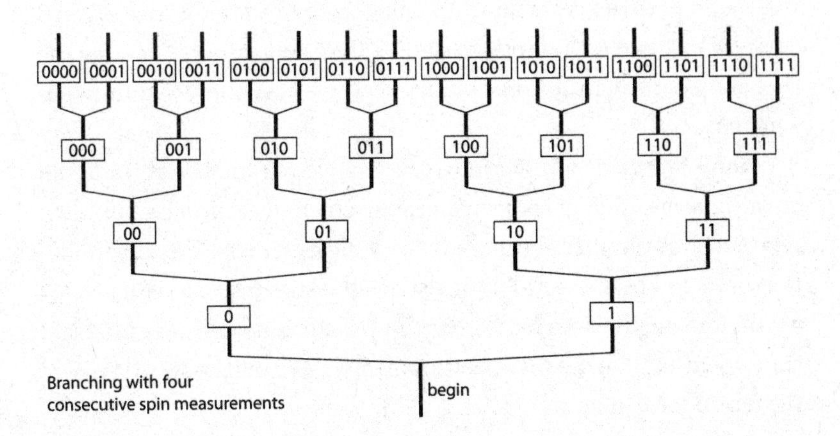

Branching with four consecutive spin measurements

If all of the copies of me in all of those different worlds stuck with the plan of including the obtained number into the text of this book, that means there are over a quadrillion different textual variations of *Something Deeply Hidden* out there in the wave function of the universe. For the most part the variations will be minor, just rearranging some 0's and 1's. But some of those poor versions of me were the unlucky ones who got all 0's or all 1's. What are they thinking right now? Probably they thought the random-number generator was broken. They certainly didn't write precisely the text I am typing at this moment.

Whatever I or the other copies of me might think about this situation, it's quite different from the frequentist paradigm for probabilities. It doesn't make too much sense to talk about the frequency in the limit of an infinite number of trials when every trial returns every result, just somewhere else in the wave function. We need to turn to another way of thinking about what probability is supposed to mean.

° ° °

Fortunately, an alternative approach to probability exists, and long predates quantum mechanics. That's the notion of *epistemic* probability, having to do with what we know rather than some hypothetical infinite number of trials.

Consider the question "What is the probability that the Philadelphia 76ers will win the 2020 NBA Championship?" (I put a high value on that personally, but fans of other teams may disagree.) This isn't the kind of event we can imagine repeating an infinite number of times; if nothing else, the basketball players would grow older, which would affect their play. The 2020 NBA Finals will happen only once, and there is a definite answer to who will win, even if we don't know what it is. But professional oddsmakers have no qualms about assigning a probability to such situations. Nor do we, in our everyday lives; we are constantly judging the likelihood of different one-shot events, from getting a job we applied for to being hungry by seven p.m. For that matter we can talk about the probability of past events, even though there is a definite thing that happened, simply because we don't know what that thing was—"I don't remember what time I left work last Thursday, but it was probably between five p.m. and six p.m., since that's usually when I head home."

What we're doing in these cases is assigning "credences"—degrees of belief—to the various propositions under consideration. Like any probability, credences must range between 0 percent and 100 percent, and your total set of credences for the possible outcomes of a specified event should add up to 100 percent. Your credence in something can change as you gather new information; you might have a degree of belief that a word is spelled a certain way, but then you go look it up and find out the right answer. Statisticians have formalized this procedure under the label of *Bayesian inference*, after Rev. Thomas Bayes, an eighteenth-century Presbyterian minister and amateur mathematician. Bayes derived an equation showing how we should update our credences when we obtain new information, and you can find his formula on posters and T-shirts in statistics departments the world over.

So there's a perfectly good notion of "probability" that applies even when something is only going to happen once, not an infinite number of times. It's a subjective notion, rather than an objective one; different people, in different states of knowledge, might assign different credences to the same outcomes for some event. That's okay, as long as everyone agrees to follow the rules about updating their credences when they learn something new. In fact, if you believe in eternalism—the future is just as real as the past; we just haven't gotten there yet—then frequentism is subsumed into Bayesianism. If you flip a random coin, the statement "The probability of the coin coming up heads is 50 percent" can be interpreted as "Given what I know about this coin and other coins, the best thing I can say about the immediate future of the coin is that it is equally likely to be heads or tails, even though there is some definite thing it will be."

It's still not obvious that basing probability on our knowledge rather than on frequencies is really a step forward. Many-Worlds is a deterministic theory, and if we know the wave function at one time and the Schrödinger equation, we can figure out everything that's going to happen. In what sense is there anything that we don't know, to which we can assign a credence given by the Born rule?

There's an answer that is tempting but wrong: that we don't know "which world we will end up in." This is wrong because it implicitly relies on a notion of personal identity that simply isn't applicable in a quantum universe.

What we're up against here is what philosophers call our "folk" understanding of the world around us, and the very different view that is suggested by modern science. The scientific view should ultimately account for our everyday experiences. But we have no right to expect that the concepts and categories that have arisen over the course of prescientific history should maintain their validity as part of our most comprehensive picture of the physical world. A good scientific theory should be compatible with our experience, but it might speak an entirely different language. The ideas we readily deploy in our day-to-day lives emerge as useful approximations of certain aspects of a more complete story.

A chair isn't an object that partakes of a Platonic essence of chairness; it's a collection of atoms arranged in a certain configuration that makes it sensible for us to include it in the category "chair." We have no trouble recognizing that the boundaries of this category are somewhat fuzzy—does a sofa count? What about a barstool? If we take something that is indubitably a chair, and remove atoms from it one by one, it gradually becomes less and less chairlike, but there's no hard-and-fast threshold that it crosses to jump suddenly from chair to non-chair. And that's okay. We have no trouble accepting this looseness in our everyday speech.

When it comes to the notion of "self," however, we're a little more protective. In our everyday experience, there's nothing very fuzzy about our self. We grow and learn, our body ages, and we interact with the world in a variety of ways. But at any one moment I have no trouble identifying a specific person that is undeniably "myself."

Quantum mechanics suggests that we're going to have to modify this story somewhat. When a spin is measured, the wave function branches via decoherence, a single world splits into two, and there are now two people where I used to be just one. It makes no sense to ask which one is "really me." Likewise, before the branching happens, it makes no sense to wonder which branch "I" will end up in. Both of them have every right to think of themselves as "me."

In a classical universe, identifying a single individual as a person aging through time is generally unproblematic. At any moment a person is a certain arrangement of atoms, but it's not the individual atoms that matter; to a large extent our atoms are replaced over time. What matters is the pattern that we form, and the continuity of that pattern, especially in the memories of the person under consideration.

The new feature of quantum mechanics is the duplication of that pattern when the wave function branches. That's no reason to panic. We just have to adjust our notion of personal identity through time to account for a situation that we never had reason to contemplate over the millennia of pre-scientific human evolution.

As stubborn as our identity is, the concept of a single person extending from birth to death was always just a useful approximation. The person you are right now is not exactly the same as the person you were a year ago, or even a second ago. Your atoms are in slightly different locations, and some of your atoms might have been exchanged for new ones. (If you're eating while reading, you might have more atoms now than you had a moment ago.) If we wanted to be more precise than usual, rather than talking about "you," we should talk about "you at 5:00 p.m.," "you at 5:01 p.m.," and so on.

The idea of a unified "you" is useful not because all of these different collections of atoms at different moments of time are literally the same, but because they are related to one another in an obvious way. They describe a real pattern. You at one moment descend from you at an earlier moment, through the evolution of the individual atoms within you and the possible addition or subtraction of a few of them. Philosophers have thought this through, of course; Derek Parfit, in particular, suggested that identity through time is a matter of one instance in your life "standing in Relation R" to another instance, where Relation R says that your future self shares psychological continuity with your past self.

The situation in Many-Worlds quantum mechanics is exactly the same way, except that now more than one person can descend from a single previous person. (Parfit would have had no problem with that, and in fact investigated analogous situations featuring duplicator machines.) Rather than talking about "you at 5:01 p.m.," we need to talk about "the person at 5:01 p.m. who descended from you at 5:00 p.m. and who ended up on the spin-up branch of the wave function," and likewise for the person on the spin-down branch.

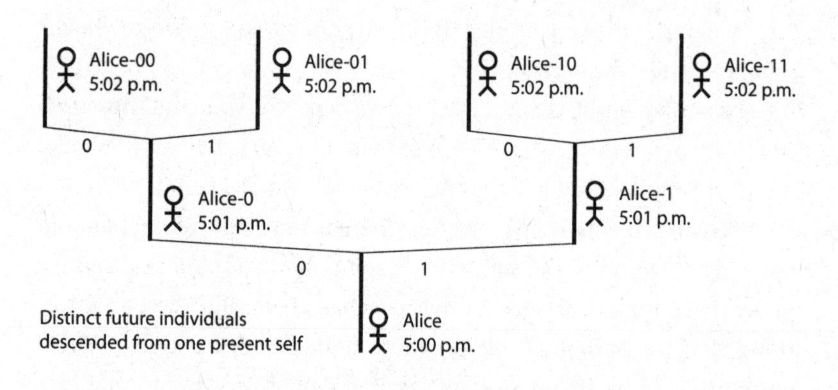

Every one of those people has a reasonable claim to being "you." None of them is wrong. Each of them is a separate person, all of whom trace their beginnings back to the same person. In Many-Worlds, the life-span of a person should be thought of as a branching tree, with multiple individuals at any one time, rather than as a single trajectory—much like a splitting amoeba. And nothing about this discussion really hinges on what we're talking about being a person rather than a rock. The world duplicates, and everything within the world goes along with it.

◦ ◦ ◦

We're now set up to confront this issue of probabilities in Many-Worlds. It might have seemed natural to think the proper question is "Which branch will I end up on?" But that's not how we should be thinking about it.

Think instead about the moment immediately after decoherence has occurred and the world has branched. Decoherence is an extraordinarily rapid process, generally taking a tiny fraction of a second to happen. From a human perspective, the wave function branches essentially instantaneously (although that's just an approximation). So the branching happens first, and we only find out about it slightly later, for example, by looking to see whether the electron went up or down when it passed through the magnetic field.

For a brief while, then, there are two copies of you, and those two copies are precisely identical. Each of them lives on a distinct branch of the wave function, but neither of them knows which one it is on.

You can see where this is going. There is nothing unknown about the wave function of the universe—it contains two branches, and we know the amplitude associated with each of them. But there is something that the actual people on these branches don't know: which branch they're on. This state of affairs, first emphasized in the quantum context by physicist Lev Vaidman, is called *self-locating uncertainty*—you know everything there is to know about the universe, except where you are within it.

That ignorance gives us an opening to talk about probabilities. In that moment after branching, both copies of you are subject to self-locating uncertainty, since they don't know which branch they're on. What they can do is assign a credence to being on one branch or the other.

What should that credence be? There are two plausible ways to go. One is that we can use the structure of quantum mechanics itself to pick out a preferred set of credences that rational observers should assign to being on various branches. If you're willing to accept that, the credences you'll end up assigning are exactly those you would get from the Born rule. The fact that the probability of a quantum measurement outcome is given by the wave function squared is just what we would expect if that probability arose from credences assigned in conditions of self-locating uncertainty. (And if you're willing to accept that and don't want to be bothered with the details, you're welcome to skip the rest of this chapter.)

But there's another school of thought, which basically denies that it makes sense to assign any definite credences at all. I can come up with all sorts of wacky rules for calculating probabilities for being on one branch of the wave function or another. Maybe I assign higher probability to being on a branch where I'm happier, or where spins are always pointing up. Philosopher David Albert has (just to highlight the arbitrariness, not because he thinks it's reasonable) suggested a "fatness measure," where the probability is proportional to the number of atoms in your body. There's no reasonable justification for doing so, but who's to stop me? The only "rational" thing to do, according to this attitude, is to admit that there's no right way to assign credences, and therefore refuse to do so.

That is a position one is allowed to take, but I don't think it's the best one. If Many-Worlds is correct, we are going to find ourselves in situations of self-locating uncertainty whether we like it or not. And if our goal is to come up with the best scientific understanding of the world, that understanding will necessarily involve an assignment of credences in these situations. After all, part of science is predicting what will be observed, even if only probabilistically. If there were an arbitrary collection of ways to assign credences, and each of them seemed just as reasonable as the other, we would be stuck. But if the structure of the theory points unmistakably to one particular way to assign such credences, and that way is in agreement with our experimental data, we should adopt it, congratulate ourselves on a job well done, and move on to other problems.

○ ○ ○

Let's say we buy into the idea that there could be a clearly best way to assign credences when we don't know which branch of the wave function we're on. Before, we mentioned that, at heart, the Born rule is just Pythagoras's theorem in action. Now we can be a little more careful and explain why that's the rational way to think about credences in the presence of self-locating uncertainty.

This is an important question, because if we didn't already know about the Born rule, we might think that amplitudes are completely irrelevant to probabilities. When you go from one branch to two, for example, why not just assign equal probability to each, since they're two separate universes? It's easy to show that this idea, known as *branch counting*, can't possibly work. But there's a more restricted version, which says that we should assign equal probabilities to branches *when they have the same amplitude*. And that, wonderfully, turns out to be all we need to show that when branches have different amplitudes, we should use the Born rule.

Let's first dispatch the wrong idea of branch counting before turning to the strategy that actually works. Consider a single electron whose vertical spin has been measured by an apparatus, so that decoherence and branching has occurred. Strictly speaking, we should keep track of the states of the apparatus, observer, and environment, but they just go along for the ride, so we won't write them explicitly. Let's imagine that the amplitudes for spin-up and spin-down aren't equal, but rather we have an unbalanced state Ψ, with unequal amplitudes for the two directions.

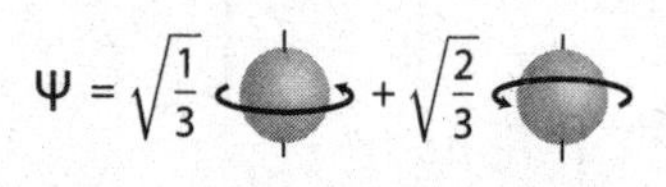

Those numbers outside the different branches are the corresponding amplitudes. Since the Born rule says the probability equals the amplitude squared, in this example we should have a 1/3 probability of seeing spin-up and a 2/3 probability of seeing spin-down.

Imagine that we didn't know about the Born rule, and were tempted to assign probabilities by simple branch counting. Think about the point of view of the observers on the two branches. From their perspective, those amplitudes are just invisible numbers multiplying their branch in the wave function of the universe. Why should they have anything to do with probabilities? Both observers are equally real, and they don't even know which branch they're on until they look. Wouldn't it be more rational, or at least more democratic, to assign them equal credences?

The obvious problem with that is that we're allowed to keep on measuring things. Imagine that we agreed ahead of time that if we measured spin-down, we would stop there, but if we measured spin-up, an automatic mechanism would quickly measure another spin. This second spin is in a state of spin-right, which we know can be written as a superposition of spin-down and spin-up. Once we've measured it (only on the branch where the first spin was up), we have three branches: one where the first spin was down, one where we got up and then down, and one where we got up twice in a row. The rule of "assign equal probability to each branch" would tell us to assign a probability of 1/3 to each of these possibilities.

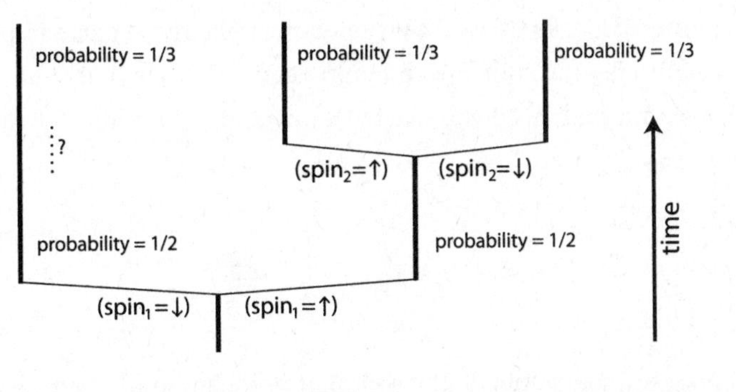

That's silly. If we followed that rule, the probability of the original spin-up branch would suddenly change when we did a measurement on the spin-down branch, going from 1/2 to 1/3. The probability of observing spin-up in our initial experiment shouldn't depend on whether someone on an entirely separate branch decides to do another experiment later on. So if we're going to assign credences in a sensible way, we'll have to be a little more sophisticated than simple branch counting.

∘ ∘ ∘

Instead of simplistically saying "Assign equal probability to each branch," let's try something more limited in scope: "Assign equal probability to branches when they have equal amplitudes." For example, a single spin in a spin-right state can be written as an equal superposition of spin-up and spin-down.

$$= \sqrt{\tfrac{1}{2}}(\ ) + \sqrt{\tfrac{1}{2}}(\ )$$

This new rule says we should give 50 percent credence to being on either the spin-up or spin-down branches, were we to observe the spin along the vertical axis. That seems reasonable, as there is a symmetry between the two choices; really, any reasonable rule should assign them equal probability.*

One nice thing about this more modest proposal is that no inconsistency arises with repeated measurements. Doing an extra measurement on one branch but not the other would leave us with branches that have unequal amplitudes again, so the rule doesn't seem to say anything at all.

But in fact it's way better than that. If we start with this simple equal-amplitudes-imply-equal-probabilities rule, and ask whether that is a special case of a more general rule that never leads to inconsistencies, we end up with a unique answer. And that answer is the Born rule: probability equals amplitude squared.

We can see this by returning to our unbalanced case, with one amplitude equal to the square root of 1/3 and the other equal to the square root of 2/3. This time we'll explicitly include a second horizontal spin-right qubit from the start. At first, this second qubit just goes along for the ride.

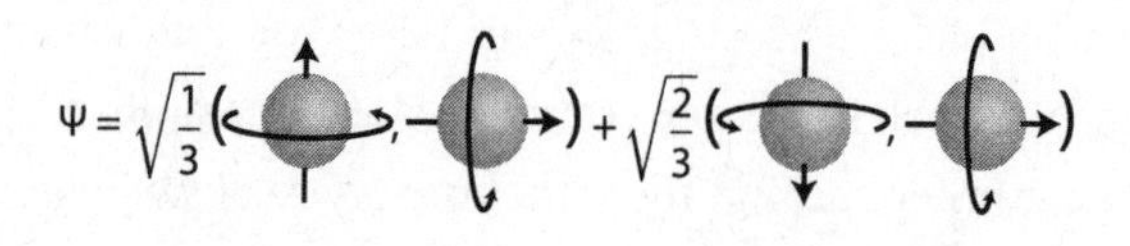

Insisting on equal probability for equal amplitudes doesn't tell us anything yet, since the amplitudes are not equal. But we can play the same game we did before, measuring the second spin along the vertical axis if the first spin is down. The wave function evolves into three

* There are more sophisticated arguments that such a rule follows from very weak assumptions. Wojciech Zurek has proposed a way of deriving such a principle, and Charles Sebens and I put forward an independent argument. We showed that this rule can be derived by insisting that the probabilities you assign for doing an experiment in your lab should be independent of the quantum state elsewhere in the universe.

components, and we can figure out what their amplitudes are by looking back at the decomposition of a spin-right state into vertical spins above. Multiplying the square root of 2/3 by the square root of 1/2 gives the square root of 1/3, so we get three branches, all with equal amplitudes.

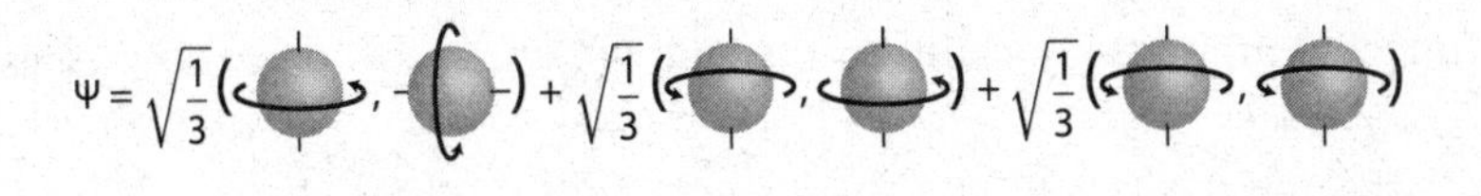

Since the amplitudes are equal, we can now safely assign them equal probabilities. Since there are three of them, that's 1/3 each. And if we don't want the probability of one branch to suddenly change when something happens on another branch, that means we should have assigned probability 1/3 to the spin-up branch even before we did the second measurement. But 1/3 is just the square of the amplitude of that branch—exactly as the Born rule would predict.

○ ○ ○

There are a couple of lingering worries here. You may object that we considered an especially simple example, where one probability was exactly twice the other one. But the same strategy works whenever we can subdivide our states into the right number of terms so that all of the amplitudes are equal in magnitude. That works whenever the amplitudes squared are all rational numbers (one integer divided by another one), and the answer is the same: probability equals amplitude squared. There are plenty of irrational numbers out there, but as a physicist if you're able to prove that something works for all rational numbers, you hand the problem to a mathematician, mumble something about "continuity," and declare that your work here is done.

We can see Pythagoras's theorem at work. It's the reason why a branch that is bigger than another branch by the square root of two can split into two branches of equal size to the other one. That's why the hard part isn't deriving the actual formula, it's providing a solid grounding for what probability means in a deterministic theory. Here we've explored one possible answer: it comes from the credences we have for being on different branches of the wave function immediately after the wave function branches.

You might worry, "But I want to know what the probability of getting a result will be even before I do the measurement, not just afterward. Before the branching, there's no uncertainty about anything—you've already told me it's not right to wonder which branch I'm going to end up on. So how do I talk about probabilities before the measurement is made?"

Never fear. You're right, imaginary interlocutor, it makes no sense to worry about which branch you'll end up on. Rather, we know with certainty that there will be two descendants of your present state, and each of them will be on a different branch. They will be identical, and they'll be uncertain as to which branch they're on, and they should assign credences given by the Born rule. But that means that all of your descendants will be in exactly the same epistemic position, assigning Born-rule probabilities. So it makes sense that you go ahead and assign those probabilities right now. We've been forced to shift the meaning of what probability is from a simple frequentist model to a more robust epistemic picture, but how we calculate things and how we act on the basis of those calculations goes through exactly as before. That's why physicists have been able to do interesting work while avoiding these subtle questions all this time.

Intuitively, this analysis suggests that the amplitudes in a quantum wave function lead to different branches having a different "weight," which is proportional to the amplitude squared. I wouldn't want to take that mental image too literally, but it provides a concrete picture that helps us make sense of probabilities, as well as of other issues like energy conservation that we'll talk about later.

$$\text{Weight of a branch} = |\text{Amplitude of that branch}|^2$$

When there are two branches with unequal amplitudes, we say that there are only two worlds, but they don't have equal weight; the one with higher amplitude counts for more. The weights of all the branches of any particular wave function always add up to one. And when one branch splits into two, we don't simply "make more universe" by duplicating the existing one; the total weight of the two new worlds is equal to that of the single world we started with, and the overall weight stays the same. Worlds get thinner as branching proceeds.

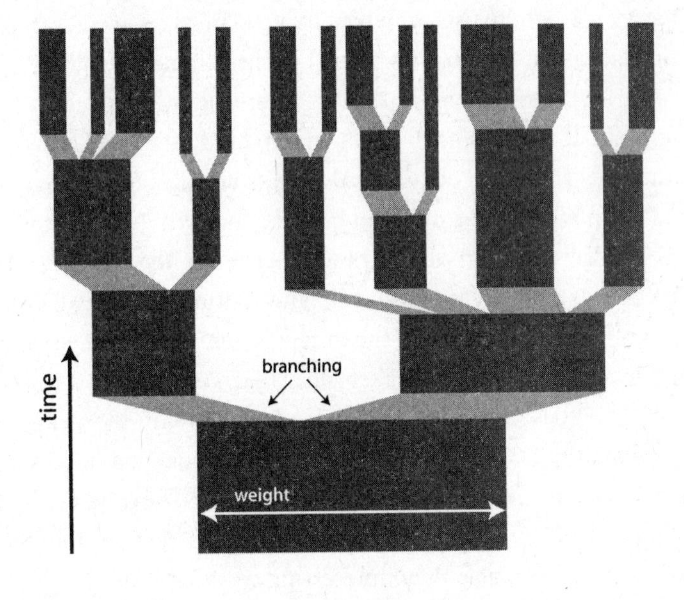

∘ ∘ ∘

This isn't the only way to derive the Born rule in the Many-Worlds theory. A strategy that is even more popular in the foundations-of-physics community appeals to decision theory—the rules by which a rational agent makes choices in an uncertain world. This approach was pioneered in 1999 by David Deutsch (one of the physicists who had been impressed by Hugh Everett at the Texas meeting in 1977), and later made more rigorous by David Wallace.

Decision theory posits that rational agents attach different amounts of value, or "utility," to different things that might happen, and then prefer to maximize the expected amount of utility—the average of all the possible outcomes, weighted by their probabilities. Given two outcomes *A* and *B*, an agent that assigns exactly twice the utility to *B* as to *A* should be indifferent between *A* happening with certainty and *B* happening with 50 percent probability. There are a bunch of reasonable-sounding axioms that any good assignment of utilities should obey; for example, if an agent prefers *A* to *B* and also prefers *B* to *C*, they should definitely prefer *A* to *C*. Anyone who goes through life violating the axioms of decision theory is deemed to be irrational, and that's that.

To use this framework in the context of Many-Worlds, we ask how a rational agent should behave, knowing that the wave function of the universe was about to branch and knowing what the amplitudes of the different branches were going to be. For example, an electron in an equal superposition of spin-up and spin-down is going to travel through a Stern-Gerlach magnet and have its spin be measured. Someone offers to pay you $2 if the result is spin-up, but only if you promise to pay them $1 if the result is spin-down. Should you take the offer? If we trust the Born rule, the answer is obviously yes, since our expected payoff is 0.5($2) + 0.5(-$1) = $0.50. But we're trying to derive the Born rule here; how are you supposed to find an answer knowing that one of your future selves will be $2 richer but another one will be $1 poorer? (Let's assume you're sufficiently well-off that gaining or losing a dollar is something you care about, but not life-changing.)

The manipulations are trickier here than in the previous case where we were explaining probabilities as credences in a situation of self-locating uncertainty, so we won't go through them explicitly, but the basic idea is the same. First we consider a case where the amplitudes on two different branches are equal, and we show that it's rational to calculate your expected value as the simple average of the two different utilities. Then suppose we have an unbalanced state like Ψ above, and I ask you to give me $1 if the spin is measured to be up and promise to give you $1 if the spin is down. By a bit of mathematical prestidigitation, we can show that your expected utility in this situation is exactly the same as if there were three possible outcomes with equal amplitudes, such that you give me $1 for one outcome and I give you $1 for the other two. In that case, the expected value is the average of the three different outcomes.

At the end of the day, a rational agent in an Everettian universe acts precisely *as if* they live in a nondeterministic universe where probabilities are given by the Born rule. Acting otherwise would be irrational, if we accept the various plausible-seeming axioms about what it means to be rational in this context.

One could stubbornly maintain that it's not good enough to show that people should act "as if" something is true; it needs to actually be true. That's missing the point a little bit. Many-Worlds quantum mechanics presents us with a dramatically different view of reality from an ordinary one-world view with truly random events. It's unsurprising that some of our most natural-seeming notions are going to have to change along with it. If we lived in the world of textbook quantum mechanics, where wave-function collapse was truly random and obeyed the Born rule, it would be rational to calculate our expected utility in a certain way. Deutsch and Wallace have shown that if we live in a deterministic Many-Worlds universe, it is rational to calculate our expected utility in exactly the same way. From this perspective, that's what it means to talk about probability: the probabilities of different events actually occurring are equivalent to the weighting we give those events when we calculate our expected utility. We should act exactly as if the probabilities we're calculating apply to a single chancy universe; but they are still real probabilities, even though the universe is a little richer than that.

Does This Ontological Commitment Make Me Look Fat?

A Socratic Dialogue on Quantum Puzzles

Alice pondered silently for a bit as she refilled her wineglass. "Let me get this straight," she said at last. "You actually want to talk about the foundations of quantum mechanics?"

"Sure," replied her father with a mischievous smile. He was a physicist himself, one who had made a successful career as a master of imposing technical calculations in particle physics. Experimentalists who smashed particles together at the Large Hadron Collider would regularly consult him on difficult questions about jets of particles created by decaying top quarks. But when it came to quantum mechanics, he was a user, not a producer. "It's about time I got a better understanding of my daughter's own research."

"Okay," she answered. In graduate school Alice had initially started down a similar career path as her father, but had gotten sidetracked by a dogged insistence on making sense of what quantum mechanics was actually saying. It seemed to her that physicists were fooling themselves by ignoring the foundations of their most important theory. A few years later, she had a PhD in theoretical physics but had landed a job as an assistant professor in the philosophy department at a major university, and was gaining a reputation as an expert on the Many-Worlds approach to quantum mechanics. “How do you want to do this?”

“I wrote down some questions,” he said as he pulled out his phone and pulled something up on its screen.

Alice felt a mixture of curiosity and trepidation. “Hit me,” she said, sniffing from the glass of Bordeaux she had poured. It was opening up nicely.

◦ ◦ ◦

“Okay,” he began. His own drink was a gin martini, not too dry, three olives. “Let’s start with the obvious. Occam’s razor. We’re all taught in kindergarten that we should prefer simple explanations over unnecessarily complicated ones. Now, if I follow your work at all—maybe I don’t—it seems to me that you’re comfortable postulating an infinite number of unseen worlds. Doesn’t that seem a bit extravagant? Directly the opposite of the simplest possible explanation?”

Alice nodded. “Well, it depends on how we define ‘simple,’ of course. My philosophy colleagues sometimes cast this as a worry about ‘ontological commitment’—roughly, the amount of stuff we need to imagine is contained in all of reality, just to describe our observed portion of it.”

“So wouldn’t Occam’s razor suggest that having too many ontological commitments is an unattractive feature in a fundamental theory?”

“Sure, but you have to be a little careful about what that commitment actually is. Many-Worlds doesn’t *assume* a large number of worlds. What it assumes is a wave function evolving according to the Schrödinger equation. The worlds are there automatically.”

Her father objected. “What do you mean by that? It’s literally called the Many-Worlds theory. Of course it assumes a large number of worlds.”

"Not really," replied Alice, becoming more animated as she warmed to the subject. "The ingredients used in Many-Worlds are ingredients that are used by *every other version of quantum mechanics.* To get rid of the other worlds, alternatives need to posit additional assumptions: either new dynamics in addition to the Schrödinger equation, or new variables in addition to the wave function, or an entirely separate view of reality. Ontologically speaking, Many-Worlds is as lean and mean as you can possibly get."

"You're kidding."

"I'm not! A much more respectable objection, to be honest, is that Many-Worlds is too lean and mean, and it's therefore a nontrivial task to map the formalism onto the messiness of our observed world."

Her father seemed to contemplate this. His cocktail sat temporarily neglected.

Alice decided to press the point. "I'll explain what I mean. If you believe that quantum mechanics is saying something about reality, you believe that an electron can be in a superposition of spin-up and spin-down, for example. And since you and I and our measuring apparatuses are made of electrons and other quantum particles, the simplest thing to assume—the thing that Occam's razor would suggest that you do—is that you and I and our measuring apparatuses can also be in superpositions, and indeed that the whole universe can be in superpositions. That is what is straightforwardly implied by the formalism of quantum mechanics, like it or not. It's certainly possible to think about complicating the theory in various ways to get rid of all those superpositions or render them unphysical, but you should imagine William of Occam looking over your shoulder, tut-tutting with disapproval."

"Seems like a bit of sophistry to me," her father grumbled. "Philosophizing aside, a bunch of in-principle-unobservable parts of your theory doesn't seem very simple at all."

"Nobody can deny that Many-Worlds involves, you know, many worlds," Alice conceded. "But that doesn't count against the simplicity of the theory. We judge theories not by the number of entities they can and do describe but by the simplicity of their underlying ideas. The idea of the integers—'–3, –2, –1, 0, 1, 2, 3 . . .'—is much simpler than the idea of, I don't know, '–342, 7, 91, a billion and three, the prime numbers less than 18, and the square root of 3.' There are more elements in the integers—an infinite number of them—but there is a simple pattern, making this infinitely big set easy to describe."

"Okay," said her father. "I can see that. There are a lot of worlds, but there is a simple principle that generates them, right? But still, by the time you actually have all those worlds, it must take an enormous amount of mathematical information to describe all them. Shouldn't we be looking for a simpler theory where they just aren't needed at all?"

"You're welcome to look," replied Alice, "and people certainly have. But by getting rid of the worlds, you end up making the theory more complicated. Think of it this way: the space of all possible wave functions, Hilbert space, is very big. It's not any bigger in Many-Worlds than in other versions of quantum theory; it's precisely the same size, and that size is more than big enough to describe a large number of parallel realities. Once you can describe superpositions of spinning electrons, you can just as easily describe superpositions of universes. If you're doing quantum mechanics at all, the potential for many worlds is there, and ordinary Schrödinger evolution tends to bring them about, like it or not. Other approaches just choose to somehow not make use of the full richness of Hilbert space. They don't want to accept the existence of other worlds, so they need to work hard to get rid of them somehow."

° ° °

"Fine," muttered her father, not fully convinced but apparently ready to move on to the next question. He took a sip of his drink and peered at his phone. "Isn't there also a philosophical problem with the theory? I'm no philosopher myself, but Karl Popper and I both know that a good scientific theory is supposed to be falsifiable. If you can't even imagine an experiment that might prove your theory wrong, it's not really science. That's exactly the situation with all these other worlds, isn't it?"

"Well, yes and no."

"That's the go-to answer to any philosophy question."

"The price we pay for being notorious sticklers for precision." Alice laughed. "Sure, Popper had this proposal that scientific theories must be falsifiable. It was an important idea. But in the back of his mind he was thinking about the difference between theories such as Einstein's general relativity, which made definite empirical predictions for the bending of light by the sun, and those like Marxist history or Freudian psychoanalysis. The problem with the latter ideas, he thought, was that no matter what actually happened, you could cook up a story to explain why it was so."

"That's what I thought. I haven't read Popper myself, but I appreciate that he put his finger on something crucial about science."

Alice nodded. "He did. But to be honest, most modern philosophers of science agree that it isn't the complete answer. Science is messier than that, and what separates science from non-science is a subtle issue."

"Everything is a subtle issue for you people! No wonder you never make any progress."

"Now, now, Dad, we are getting at something significant here. What Popper was ultimately trying to pinpoint is that a good scientific theory has two characteristics. First, it is *definite*: you can't just twist the theory to 'explain' anything at all, as Popper feared you could do with dialectical materialism or psychoanalysis. Second, it is *empirical*: theories are not deemed true by sheer reason alone. Rather, one imagines many different possible ways the world could be, each corresponding to a different theory, and then one chooses among the theories by going out and actually looking at the world."

"Exactly." Her father seemed to think that the advantage was his on this one. "Empirical! But if you can't actually observe those worlds, there's nothing really empirical about your theory at all."

"Au contraire," Alice replied. "Many-Worlds embodies both of these features perfectly. It is not a just-so story that can be adapted to any observed set of facts. Its postulates are simple: the world is described by a quantum wave function that evolves according to the Schrödinger equation. Those postulates are eminently falsifiable. Just do an experiment showing that quantum interference doesn't occur when it should, or that entanglement really can be used for superluminal communication, or that a wave function really does collapse even without decoherence. Many-Worlds is the most falsifiable theory ever invented."

"But those aren't tests of Many-Worlds," her father protested, unwilling to concede ground on this one. "Those are just tests of quantum mechanics generally."

"Right! But Everettian quantum mechanics is just pure, austere quantum mechanics without any additional ad hoc assumptions. If you do want to introduce extra assumptions, then by all means we can ask whether those new assumptions are testable."

"Come now. The defining feature of Many-Worlds is the existence of all those worlds out there. Our world can't interact with them, so that particular aspect of the theory is untestable."

"So what? Every good theory makes some predictions that are untestable. Our current theoretical understanding of general relativity predicts that the force of gravity will not tomorrow suddenly turn off for a period of one millisecond in a particular region of space ten meters across and twenty million light-years away. That's a completely untestable prediction, of course, but we maintain a very high credence that it's true. There's no reason for gravity to behave in that way, and imagining that it did leaves us with a much uglier theory than the one we have. The additional worlds in Everettian quantum mechanics have exactly this character: they are inescapable predictions of a simple theoretical formalism. We should accept them unless we have a specific reason not to.

"And besides," Alice rushed on, "the other worlds could be detected in principle, if we got incredibly lucky. They haven't gone away, they're still there in the wave function. Decoherence makes it fantastically unlikely for one world to interfere with another, but not metaphysically impossible. I wouldn't suggest applying for grant money to do such an experiment, though; it would be like mixing cream into coffee and waiting around for them to spontaneously unmix themselves."

"Don't worry, I wasn't planning on it. I just don't think Karl Popper would be very happy with your approach to the philosophy of science."

"I've got you there, Dad," said Alice. "Popper himself was a harsh critic of the Copenhagen interpretation, which he called a 'mistaken and even a vicious doctrine.' In contrast, he had good things to say about Many-Worlds, which he accurately described as 'a completely objective discussion of quantum mechanics.'"

"Seriously? Popper was an Everettian?"

"Well, no," Alice admitted. "He ultimately parted ways with Everett because he couldn't understand why the wave function would branch but branches wouldn't later fuse back together. I mean, that's a good question, but it's one we can answer."

"I'm sure you can. Where did he come down on the foundations of quantum mechanics?"

"He developed his own formulation of quantum mechanics, but it never really caught on."

"Ha! Philosophers."

"Yeah. We're better at telling you why your theory is wrong than at proposing better ones."

○ ○ ○

Alice's father sighed. "Fine. I'm not saying you're convincing me of anything, but I don't want to get bogged down in philosophical hair-splitting. Now that you mention it, Popper's question does seem kind of reasonable. Why don't worlds fuse together as well as branch apart? If we have a spin that is an equal superposition of up and down, we can predict the probability of observing either outcome if we do a measurement in the future. But if we have a spin that is purely up, and we are told that it was just measured, we have absolutely no way of knowing what kind of superposition it was in pre-measurement (except that it wasn't purely down). Where does the difference come from?"

Alice seemed ready for this one. "That's just thermodynamics, really. Or at least, it's the arrow of time, pointing from the past to the future. We remember yesterday but not tomorrow; cream and coffee mix together but they don't spontaneously unmix. Wave functions branch, but don't unbranch."

"Sounds suspiciously circular. As I understand it, one of the purported features of Many-Worlds is that wave functions only obey the Schrödinger equation; there's no separate collapse postulate. Back when I learned quantum mechanics, we knew that wave functions collapsed toward the future and not toward the past, and that was part of the assumptions. I don't see why that should still be true for Everett, where the Schrödinger equation is completely reversible. What do cream and coffee have to do with wave functions?"

Alice nodded. "Perfectly good question. Let's set the stage a bit. The second law of thermodynamics posits that entropy—roughly, the disorderliness or randomness of a configuration, as you know—never decreases in closed systems. Ludwig Boltzmann explained this back in the 1870s. Entropy counts the number of ways that atoms can be arranged so that the system looks the same from a macroscopic perspective. The reason why it increases is simply that there are many more ways to be high-entropy than to be low-entropy, so it's improbable that entropy would ever go down. Right?"

"Sure," her father agreed. "But that's all classical; Boltzmann didn't know anything about quantum mechanics."

"Right, but the basic idea is the same. Boltzmann explained why entropy tends to increase, but he didn't give a reason why it was ever low in the first place. These days we appreciate that it is a cosmological fact that the universe started out right after the Big Bang in an orderly state, and entropy has naturally been increasing ever since, and so we have time's arrow. We don't really know why the early universe had such a low entropy, though some of us have ideas."

"And this is relevant because . . ."

"Because for Everettians, the explanation of the quantum arrow of time is the same as that of the entropic arrow of time: the initial conditions of the universe. Branching happens when systems become entangled with the environment and decohere, which unfolds as time moves toward the future, not the past. The number of branches of the wave function, just like the entropy, only increases with time. That means that the number of branches was relatively small to begin with. In other words, that there was a relatively low amount of entanglement between various systems and the environment in the far past. As with entropy, this is an initial condition we impose on the state of the universe, and at the present time we don't know for sure why it was the case."

"Okay," said her father. "It's good to admit what we don't know. We explain the arrow of time, at least according to the current state of the art, by appeal to special initial conditions in the past. Is it a single condition that explains both the thermodynamic arrow and the quantum arrow, or is that just an analogy?"

"I think it's more than an analogy, but to be honest, this is a subject that could probably use a bit more rigorous investigation," Alice replied. "There certainly seems to be a connection. Entropy is related to our ignorance. If a system has low entropy, there are relatively few microscopic configurations that would look that way, so we know a lot about it just from its macroscopically observable features; if it has high entropy, we know relatively little. John von Neumann realized that we can say something similar about entangled quantum systems. If a system is completely unentangled with anything else, we can safely talk about its wave function in isolation from the rest of the world. But when it is entangled, the individual wave function is undefined, and we can only talk about the wave function for the combined system."

Her father brightened. "Von Neumann was a brilliant guy, a real hero. There were an amazing number of Hungarian physicists who emigrated to the US—Szilard, Wigner, Teller—but he was the top. I do vaguely remember that he derived a formula for entropy."

Alice agreed. "No question. Von Neumann realized that there was a mathematical equivalence between a classical situation when we're unsure about the exact state of a system, which gives rise to entropy, and the quantum situation where two subsystems are entangled, so we can't talk about the wave function of either piece separately. He derived a formula for the 'entanglement entropy' of a quantum system. The more entangled something is with the rest of the world, the higher its entropy."

"Aha," exclaimed her father excitedly. "I see where you're going with this. The fact that wave functions only branch forward in time and not backward is not simply reminiscent of the fact that entropy increases—it's the same fact. The low entropy of the early universe corresponds to the idea that there were many unentangled subsystems back then. As they interact with each other and become entangled, we see that as branching of the wave function."

"Exactly," Alice responded, with something like daughterly pride. "We're still not sure why the universe is like that, but once we accept that the early universe was in a relatively unentangled, low-entropy state, everything else follows."

"But wait a minute." Her father seemed to have just realized something. "According to Boltzmann, entropy is only likely to increase, it's not an absolute rule. It's ultimately due to the random motions of atoms and molecules, so there's a nonzero probability that entropy will spontaneously go down. Does that mean that it's possible that decoherence will someday reverse, and worlds actually will fuse together rather than branching apart?"

"Absolutely," said Alice with a nod. "But just like with entropy, the chance of that happening is so preposterously small that it's irrelevant to our daily lives, or to any experiment in the history of physics. It's extremely unlikely that two macroscopically distinct configurations have recohered even once in the lifetime of our universe."

"So you're saying there's a chance?"

"I'm saying that if your worry about Many-Worlds is that branches of the wave function will someday come back together, you've clearly exhausted all the reasonable worries and are grasping at straws."

◦ ◦ ◦

"Well, let's not get too full of ourselves just yet," her father muttered, seemingly returning to his skeptical stance. He lifted the toothpick from his glass and bit off an olive. "Let me try to understand what the theory actually says. Is it right to say that the number of worlds being produced at every moment is literally infinitely big?"

"Well," replied Alice, somewhat tentatively, "I'm afraid an honest answer to that question is going to require a bit more philosophical hair-splitting."

"Why am I not surprised?"

"We can go back to entropy as an analogy. When Boltzmann came up with his entropy formula, he counted the number of microscopic arrangements of a system that looked macroscopically the same. From there, he was able to argue that entropy should naturally increase."

"Sure," said her father. "But that is real, honest physics, something we can test experimentally. Not sure what it has to do with your Many-Worlds flights of fancy."

"We say that now. But you have to imagine what people were thinking back at the time." Alice was settling comfortably into professor mode, her Bordeaux momentarily forgotten. "Boltzmann was right, but a number of objections were raised to his idea. One was that he was turning entropy from an objective feature of a physical system into a subjective one, which depended on some notion of 'looks the same.' Another was that he demoted the second law from an absolute statement to a mere tendency—it wasn't that entropy necessarily increased, it was just very likely to do so. Particles jiggle around randomly, and it's extremely probable that they will evolve toward a higher-entropy state, but it's not a lawlike certainty. With the wisdom of accumulated years, we can see that the subjective nature of Boltzmann's definition does not stop it from being a useful one, and the fact that the second law is a really good approximation rather than an absolute unbreakable law is more than good enough for whatever purposes we may have."

"I get that," answered her father. "Entropy is an objectively real thing, but we can define and measure it only after making a few decisions. But that never really bothered me—it's useful! I'm not sure that extra worlds really are."

"We'll get there, but first let me elaborate on this analogy. Like entropy, the notion of a 'world' in Everettian quantum mechanics is a higher-level concept, not a fundamental one. It's a useful approximation that provides genuine physical insight. The separate branches of the wave function aren't put in as part of the basic architecture of the theory. It's just extraordinarily convenient for us human beings to think of a superposition of many such worlds, rather than treating the quantum state as an undifferentiated abstraction."

Her father's eyes widened a bit. "This is worse than I feared. It sounds like you're going to tell me that a 'world' isn't even a well-defined concept in Many-Worlds."

"They're just as well defined as entropy is. If we were a nineteenth-century Laplace demon, who knew the position and momentum of every particle in the universe, we would never have to stoop to defining a coarse-grained notion like 'entropy.' Likewise, if we knew the exact wave function of the universe, we would never have to talk about 'branches.' But in both cases we are poor finite creatures with dramatically incomplete information, and invoking these higher-level concepts is extremely useful."

Alice could tell that her father was losing patience. "I just want to know how many worlds there are," he said. "If you can't answer that, you're not doing a very good sales job here."

"Must be that devotion to honesty under any circumstances that you inculcated into me at a young age," Alice said with a shrug. "It depends on how we divide the quantum state into worlds."

"And isn't there some obvious right way?"

"Sometimes! In simple situations where measurements have a manifestly discrete outcome, like measuring the spin of an electron, we can safely say that the wave function branches in two, and the number of worlds (whatever that was) doubles. When we're measuring a quantity that is in principle continuous, like the position of a particle, things are less well defined. In that case we can define a total weight attached to a certain range of outcomes, the wave function squared, but not an absolute number of branches. That number would depend on how finely we want to subdivide our description of the measurement outcome, which is ultimately a choice that's up to us. One of my favorite quotes along these lines is from David Wallace: 'Asking how many worlds there are is like asking how many experiences you had yesterday, or how many regrets a repentant criminal has had. It makes perfect sense to say that you had many experiences or that he had many regrets; it makes perfect sense to list the most important categories of either; but it is a nonquestion to ask *how many*.'"

Alice's father didn't really seem satisfied by this. After a thoughtful pause, he responded, "Look, I'm trying to be fair here. I'll accept that the worlds are not fundamental, so there is something approximate about how they are defined. But surely you can tell me whether there are just a finite number of them or the number is truly infinite."

"It's a fair question," Alice agreed, maybe a bit reluctantly. "Unfortunately, we don't know the answer. There's an upper limit to the number of worlds, which is just the size of Hilbert space, the space of all possible wave functions."

"But we know that Hilbert space is infinitely big," interjected her father. "Even for just one particle, Hilbert space is infinite-dimensional, not to mention for quantum field theory. So the number of worlds sounds like it's infinite."

"We're not sure whether the Hilbert space for our actual universe has a finite or infinite number of dimensions. We certainly know of some systems for which the appropriate Hilbert space is finite-dimensional. A single qubit is either spin-up or spin-down, so it corresponds to a two-dimensional Hilbert space. If we have N qubits, the corresponding Hilbert space is 2^N-dimensional—the size of Hilbert space grows exponentially as we include more particles. A cup of coffee contains roughly 10^{25} electrons, protons, and neutrons, each of whose spins is described by a qubit. So the Hilbert space for a cup of coffee—just including the spins, not yet worrying about the locations of the particles—has a dimensionality of about $2^{10^{25}}$.

"Needless to say," continued Alice, "that's a crazy-big number. One followed by 10^{25} zeroes, if you wrote it in binary. Which you wouldn't have time to do, even if you had been working for the entire lifetime of our observable universe."

"But you're obviously cheating, the real number is much bigger than that," said her father. "You're counting spins, but real particles have locations in space too. And there are an infinite number of such locations. That's why the Hilbert space for a collection of particles is infinite-dimensional—the number of dimensions is just the number of possible measurement outcomes."

"Right. And it's true, Hugh Everett himself thought that every quantum measurement split the universe into an infinite number of worlds, and he was comfortable with that. Infinity sounds like a big number, but we use infinite quantities in physics all the time. The number of real numbers between 0 and 1 is infinite, as you know. If Hilbert space is infinite-dimensional, it doesn't make much sense to talk about the number of individual worlds. But we can group a set of similar worlds together, and talk about the total weight (amplitude-squared) they have compared to some other group."

"Great. So Hilbert space is infinite-dimensional, and the number of worlds is infinite, but you want to claim that we should only talk about the relative weight of different kinds of worlds?"

"No, I'm not done yet," Alice insisted. "The real world isn't a bunch of particles, nor is it even described by quantum field theory."

"It's not?" said her father in mock dismay. "What have I been doing all my life?"

"You've been ignoring gravity," replied Alice, "which is a perfectly sensible thing to do while you're thinking about particle physics. But there are indications from quantum gravity that the number of distinct possible quantum states is finite, not infinite. If that's true, there is a maximum number of worlds we could sensibly talk about, given by the dimensionality of Hilbert space. The kinds of estimates that get thrown around for the number of dimensions of the Hilbert space of our observable universe are things like $2^{10^{122}}$. A big number," Alice admitted, "but even very big finite numbers are much smaller than infinity."

Her father seemed to think about this. "Huh. I'm not really sure we know anything very reliable about quantum gravity—"

"Maybe we don't. That's why I said we really don't know if the number of worlds is finite or infinite."

"Fair enough. But that raises a totally new worry. It seems to me that branching should be happening all the time, every time a quantum system becomes entangled with its environment. Is it conceivable that this number you just quoted, while mind-bogglingly large, isn't large enough? Are we sure there's enough room in Hilbert space for all the branches of the wave function that are being produced as the universe evolves?"

"Hmm, I never thought about that, to be honest." Alice grabbed a napkin and started scribbling some numbers on it. "Let's see, there are about 10^{88} particles within our observable universe, mostly photons and neutrinos. For the most part these particles travel peacefully through space, not interacting or becoming entangled with anything. So as a generous overestimate, let's imagine that every particle in the universe interacts and splits the wave function in two a million times per second, and has been doing so since the Big Bang, which was about 10^{18} seconds ago. That's $10^{88} \times 10^{6} \times 10^{18} = 10^{112}$ splittings, producing a total number of branches of $2^{10^{112}}$.

"Nice!" Alice seemed pleased with herself. "That's still a really big number, but it's much smaller than the number of dimensions in the Hilbert space of the universe. Pitifully smaller, really. And it should be a safe overestimate of the number of branches required. So even if the question of how many branches there are doesn't have a definite answer, we don't need to worry that Hilbert space is going to run out of room."

∘ ∘ ∘

"Well, good, I was worried there for a second." Her father's martini tasted pleasantly briny from the olives. He regarded Alice, a glint in his eye. "Had you really never asked yourself that question before?"

"I think most Everettians train themselves to think of the relative weights of various different branches of the wave function, rather than actually counting anything. We don't know the ultimate answer, so it doesn't seem too fruitful to worry about it."

"I'll have to process this a bit, because I always thought that there were supposed to be an infinite number of worlds, and that Many-Worlds implied that everything happened somewhere. That every possible world exists out there in the wave function. I thought that was the selling point. When I was stuck on a calculation, it was comforting to think that there was another world in which I was a llama, or a genius billionaire playboy philanthropist."

"Wait, you're not?" Alice feigned surprise. "I always thought you looked a bit like a llama."

"I mean, for that matter, in some world I should be a billionaire llama."

"Before we get off track," she continued, "let me just note that it's not 'you' who would be a llama or a billionaire, those would be other beings entirely. I'm sure we'll come back to that. But of more direct relevance to the issue, Many-Worlds doesn't say 'everything possible happens'; it says 'the wave function evolves according to the Schrödinger equation.' Some things don't happen, because the Schrödinger equation never leads to them happening. For example, we will never see an electron spontaneously convert into a proton. That would change the amount of electric charge, and charge is strictly conserved. So branching will never create, for example, universes with more or less charge than we started with. Just because many things happen in Everettian quantum mechanics doesn't mean that everything does."

Alice's father raised his eyebrows in skepticism. "Dear, you are surely nitpicking to save face. Maybe not strictly everything happens, but I believe it's true that a great many crazy-sounding things do happen in various worlds, no?"

"Sure, I'm happy to admit that. Every time you run into a wall, the wave function branches into a number of worlds: some where you injure your nose, some where you harmlessly tunnel right through, and others where you bounce off and are thrown across the room, for example."

"But that matters a lot, doesn't it? In ordinary quantum mechanics the probability of a macroscopic object tunneling through a wall is not zero, but it's unimaginably tiny, and we can just ignore it. In Many-Worlds, the probability is 100 percent that it happens in some world."

Alice nodded, but her expression was that of someone who had gone over this ground many times before. "You're absolutely right that this is a difference. But I would argue that it doesn't matter a single bit. If you accept how Everettians derive the Born rule, you should *act as if* there is a probability of you tunneling through the wall, and that probability is so preposterously small that there's no reason whatsoever to take it into consideration as you go through your everyday life. And if you don't accept that argument, there is a much more serious worry about Many-Worlds for you to fret over."

Her father was determined. "I think the issue of these low-probability worlds is important. What about those observers who, in the ensemble of Everettian worlds, end up seeing events that seemingly defy our Born rule predictions? If we measure a spin fifty times, there will be branches on which all of the results read spin-up, and others on which they all read spin-down. What are those poor observers supposed to conclude about quantum mechanics?"

"Well," said Alice, "mostly we have to say, too bad for them. Stuff happens. But the total weight assigned to such observers is so small that we shouldn't worry about them too much. Not to mention that, after they get fifty spin-ups in a row, the next fifty trials will still map onto the Born-rule predictions with overwhelming probability. Most likely they will attribute their original lucky streak to experimental error, and have a fun story to tell their lab mates. It's just like a classical universe that is just really big. If conditions that we see in the universe around us continue infinitely far in every direction, it is overwhelmingly probable that there are other civilizations just like ours—an infinite number, in fact—doing experiments to test quantum mechanics. Even if each of them is likely to see Born-rule probabilities, given that there are an infinite number of them, some of them will see very different statistics. In that case they may be led to draw incorrect conclusions about how quantum mechanics works. Those observers would be unlucky, but we can take consolation in the fact that they are also very infrequent among the set of all observers in the universe."

"Small consolation for them! In your view of physics, there will always be observers out there who get the laws of nature utterly wrong."

"Nobody ever promised them a rose garden. That worry exists in any theory where there are sufficiently large numbers of observers; Many-Worlds is just one example of such a theory. The point is that in Everettian quantum mechanics, there is a way to compare all of the different worlds: take the amplitudes of their branches, and square them. The branches in which very surprising things happen have very, very tiny amplitudes. They are rare in the set of all worlds. We shouldn't be any more bothered by their existence than we are by unlucky observers in infinitely large universes."

° ° °

"Not sure I'm convinced here, but let's just enter my worry into the record and move on." He squinted at the list of questions he had brought up on his phone. "I've been doing a bit of reading—even some of your papers—and one thing I do appreciate about Many-Worlds is that it removes any lingering mystery about when a measurement takes place. There's nothing special about measurement; it's just when a quantum system that's in a superposition becomes entangled with the larger environment, leading to decoherence and branching of the wave function. But there is only one wave function, the wave function of the universe, which describes everything throughout space. How should we think about branching from a global perspective? Does branching happen all at once, or does it gradually spread out from the system where the interaction occurred?"

"Oh boy. I have a feeling this is going to be another unsatisfying answer." Alice paused to slice off a piece of cheese. She carefully arranged it on a cracker as she thought about the best response. "Basically: that's up to you. Or, to put the point in more respectable-sounding language, the very phenomenon of 'branching' is one that we humans invent to provide a convenient description of a complicated wave function, and whether we think of branching as happening all at once or as spreading out from a point depends on what's more convenient for the situation."

Her father shook his head. "I thought branching was the whole point. How can you hold up Many-Worlds as a respectable scientific theory if not only can't you observe the other branches, and not only can't you count them, but you don't even have a definite criterion for how it happens? Branching is just, like, your opinion, man?" He had always been just a little too fond of movie references.

"In a sense, sure. But there are better and worse opinions to have. You may prefer a description in which nothing travels faster than the speed of light. What actually matters is that you can't communicate or send information faster than light, and that's true no matter what description you choose to use. But if it makes you feel better to limit an apparently physical effect like branching to propagate no faster than light, you are perfectly welcome to do that. In that case, the number of branches of the wave function would be different depending on where you were in spacetime." She took out a fresh napkin and began scribbling again, this time making little diagrams out of straight lines. "Here we have space going from left to right and time going upward. Light beams that could potentially be emitted from an event will move upward at forty-five-degree angles. If we start with just a single branch of the wave function, we can imagine branching happening at that event, and then propagating upward in time, but only growing at the speed of light. Observers farther away would be described by a single branch, while nearer ones would be described by two branches. This fits well with the idea that distant observers have no way of knowing, or being influenced by, the branching event, while those nearby do."

Her father studied the diagram. "I see. I guess I assumed that branching happened simultaneously throughout the universe, which bothered me as someone who is quite fond of special relativity. I'm sure you know as well as I do that different observers will define simultaneity differently. I kind of like this picture better, where branching propagates outward at the speed of light. All the effects look pretty local."

Alice waved her hands before she resumed drawing. "But the other way works too. We are equally allowed to describe branching as happening all throughout the universe all at once. This view is helpful when we derive the Born rule using self-locating uncertainty, as we can sensibly talk about which branch you are on immediately after the branching occurs, no matter where it happened. Because of relativity, observers moving at different speeds will draw the branches differently, but there's no observational difference caused by doing so."

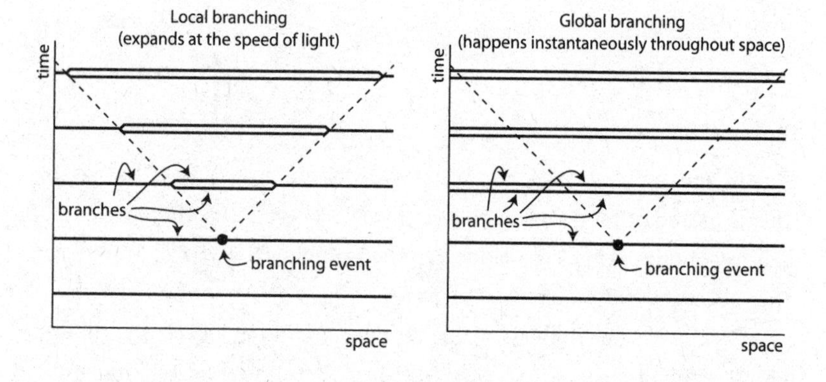

"Arrgh! You've just undone all of your good work. Now you're telling me that branching can just as well be thought of as completely nonlocal."

"Yeah, but what I'm actually saying is that the question 'Is Many-Worlds a local theory?' isn't quite the right one to ask. It would be better to ask, 'Can we describe branching as a local process, proceeding only inside the future light cone of an event?' The answer is 'Yes, but we can equally well describe it as a nonlocal process, occurring instantly throughout the universe.'"

Her father put his hands over his face, but he seemed to be trying to absorb this, not just giving up in frustration. Then he got up and mixed himself another martini, brow furrowed. He returned to his seat, drink in one hand and some peanuts in the other. "I guess the point is that whether or not I think a person far away has branched, it doesn't make any difference to them. I can think of them as being just one copy, or as two copies that are absolutely identical. It's just a matter of description."

"Exactly!" Alice exclaimed. "Whether we think about branching as propagating outward at the speed of light or happening all at once is just a question of what's most convenient. It's no more worrisome than the fact that we can measure length in centimeters or inches."

Her father rolled his eyes. "What kind of barbarian measures length in inches?"

◦ ◦ ◦

"Okay, let's shift gears," he said after a moment. "I know that string theorists and other people who aren't very tethered to reality are fond of talking about extra dimensions. Do the branches live there? Where are these other worlds located, anyway?"

"Oh, come on, Robert." Alice tended to call her father by his first name when she was annoyed with him. "You know better than that. The branches aren't 'located' anywhere. If you're stuck thinking of things as having locations in space, it might seem natural to ask about where the other worlds are. But there is no 'place' where those branches are hiding; they simply exist simultaneously, along with our own, effectively out of contact with it. I suppose they exist in Hilbert space, but that's not really a 'place.' There are more things in heaven and earth than are dreamt of in your philosophy." She was proud to keep her references Shakespearean.

"Yeah, I know. We're a couple of drinks in, I thought I should toss you a softball."

° ° °

He scrolled down the document on his phone a bit. "All right, let's get more serious here. This one has been bugging me forever. What about conservation of energy? Where does all that *stuff* come from when you suddenly create a whole new universe?"

"Well," replied Alice, "just think about ordinary textbook quantum mechanics. Given a quantum state, we can calculate the total energy it describes. As long as the wave function evolves strictly according to the Schrödinger equation, that energy is exactly conserved, right?"

"Sure."

"That's it. In Many-Worlds, the wave function obeys the Schrödinger equation, which conserves energy."

"But what about the extra worlds?" her father insisted. "I could measure the energy contained in this world I see around me, and you say it's being duplicated all the time."

Alice felt she was on firm ground with this one. "Not all worlds are created equal. Think about the wave function. When it describes multiple branched worlds, we can calculate the total amount of energy by adding up the amount of energy in each world, times the weight (the amplitude squared) for that world. When one world divides in two, the energy in each world is basically the same as it previously was in the single world (as far as anyone living inside is concerned), but their contributions to the total energy of the wave function of the universe have divided in half, since their amplitudes have decreased. Each world got a bit thinner, although its inhabitants can't tell any difference."

"Mathematically I see what you're saying," admitted her father. "But I seem to be lacking some intuition here. I have, say, a bowling ball, with a certain mass and potential energy. But then someone in the next room observes a quantum spin and branches the wave function. Now there are two bowling balls, each of which has the energy of the previous one. No?"

"That ignores the amplitudes of the branches. The contribution of the bowling ball to the energy of the universe isn't just the mass and the potential energy of the ball; it's that, times the weight of its branch of the wave function. After the splitting it looks like you have two bowling balls, but together they contribute exactly as much to the energy of the wave function as the single bowling ball did before."

Her father seemed to ponder this. "I'm not sure I agree with you, but I think you're wearing me down," he muttered. After a moment he turned back to his list of questions.

◦ ◦ ◦

"You know, I think I only have one question left." Alice's father put away his phone, drank some more of his second martini, and leaned in a bit. "Do you really believe this? Honestly? That multiple copies of me come into existence every time someone measures the spin of a particle?"

Alice sat back in her chair, savored a bit of her wine, and looked thoughtful. "You know, I really do. At least, I personally find Everettian quantum mechanics, and all the many worlds that it implies, to be by far the most plausible version of quantum theory that I know of. If that means I must accept that my present self will evolve into a number of slightly different future selves who will never be able to talk to each other, I'm willing to accept that. Subject, as always, to being updated in the future if new information comes along, either in the form of experimental results or new theoretical insights."

"Such a good empiricist." Her father smiled.

"Let me quote David Deutsch," Alice offered. "He once said, 'Despite the unrivaled empirical success of quantum theory, the very suggestion that it may be *literally true as a description of nature* is still greeted with cynicism, incomprehension, and even anger.'"

"What's that supposed to mean? Every physicist thinks quantum mechanics describes nature."

"I think when Deutsch says 'quantum theory,' he implicitly means Many-Worlds." Now it was Alice's turn to smile. "What he was getting at was that many people reject Everettian quantum mechanics more out of a visceral sense of distaste than a principled set of worries. But as philosopher David Lewis once put it, 'I do not know how to refute an incredulous stare.'"

"I hope you're not including me there." Alice's father looked slightly affronted. "I've just been trying to understand the theory in a principled way."

"You have!" Alice replied. "The conversation we've just been having—whether or not I convinced you of anything at all, this is what all thoughtful physicists should be talking about. What matters to me is not that everyone become an Everettian, but that people take the challenge of understanding quantum mechanics seriously. I'd much rather have a dialogue with someone who is a dedicated proponent of hidden variables, for example, than try to engage the interest of someone who just doesn't care."

Her father nodded. "It's taken me a while, I admit. But yes, I do care." He smiled at his daughter. "Our mission is to understand things, isn't it?"

Other Ways

Alternatives to Many-Worlds

David Albert, now a philosophy professor at Columbia and one of the world's leading researchers in the foundations of quantum mechanics, had a very typical experience as a graduate student who became interested in quantum foundations. He was in the PhD program in the physics department at Rockefeller University when, after reading a book by eighteenth-century philosopher David Hume on the relationship of knowledge and experience, he came to believe that what physics lacked was a good understanding of the quantum measurement problem. (Hume didn't know about the measurement problem, but Albert connected dots in his head.) Nobody at Rockefeller in the late 1970s was interested in thinking along those lines, so Albert struck up a long-distance collaboration with the famous Israeli physicist Yakir Aharonov, resulting in several influential papers. But when he suggested submitting that work for his PhD thesis, the powers that be at Rockefeller were aghast. Under penalty of being kicked out of the program entirely, Albert was forced to write a separate thesis in mathematical physics. It was, as he recalled, "clearly being assigned because it was thought it would be good for my character. There was an explicitly punitive element there."

Physicists have been very bad at coming to consensus about what the foundations of quantum mechanics actually are. But in the second half of the twentieth century, they did come to a remarkable degree of consensus on a related issue: whatever the foundations of quantum mechanics are, we certainly shouldn't *talk* about them. Not while there was real work to be done, doing calculations and constructing new models of particles and fields.

Everett, of course, left academia without even trying to become a physics professor. David Bohm, who had studied and worked under Robert Oppenheimer in the 1940s, proposed an ingenious way of using hidden variables to address the measurement problem. But after a seminar in which another physicist explained Bohm's ideas, Oppenheimer scoffed out loud, "If we cannot disprove Bohm, then we must agree to ignore him." John Bell, who did more than anyone to illuminate the apparently nonlocal nature of quantum entanglement, purposefully hid his work on this subject from his colleagues at CERN, to whom he appeared as a relatively conventional particle theorist. Hans Dieter Zeh, who pioneered the concept of decoherence as a young researcher in the 1970s, was warned by his mentor that working on this subject would destroy his academic career. Indeed, he found it very difficult to publish his early papers, being told by journal referees that "the paper is completely senseless" and "quantum theory does not apply to macroscopic objects." Dutch physicist Samuel Goudsmit, serving as the editor of *Physical Review*, put out a memo in 1973 explicitly banning the journal from even considering papers on quantum foundations unless they made new experimental predictions. (Had that policy been in place earlier, the journal would have had to reject the Einstein-Podolsky-Rosen paper, as well as Bohr's reply.)

Yet, as these very stories make clear, despite a variety of hurdles put up in their way, a subset of physicists and philosophers nevertheless persevered in the effort to better understand the nature of quantum reality. The Many-Worlds theory, especially once the process of wave-function branching has been illuminated by decoherence, is one promising approach to answering the puzzles raised by the measurement problem. But there are others worth considering. They are worthwhile both because they might actually be right (which is always the best reason) and also because comparing the very different ways in which they work helps us to better appreciate quantum mechanics, no matter what our personal favorite approach happens to be.

An impressive number of alternative formulations of quantum theory have been proposed over the years. (The relevant Wikipedia article lists sixteen "interpretations" explicitly, along with a category for "other.") Here we'll consider three basic competitors to the Everett approach: dynamical collapse, hidden variables, and epistemic theories. While far from comprehensive, these serve to illustrate the basic strategies that people have taken.

o o o

The virtue of Many-Worlds is in the simplicity of its basic formulation: there is a wave function that evolves according to the Schrödinger equation. All else is commentary. Some of that commentary, such as the split into systems and their environment, decoherence, and branching of the wave function, is extremely useful, and indeed indispensable to matching the crisp elegance of the underlying formalism to our messy experience of the world.

Whatever your feelings might be about Many-Worlds, its simplicity provides a good starting point for considering alternatives. If you remain profoundly skeptical that there are good answers to the problem of probability, or are simply repulsed by the idea of all those worlds out there, the task you face is to modify Many-Worlds in some way. Given that Many-Worlds is just "wave functions and the Schrödinger equation," a few plausible ways forward immediately suggest themselves: altering the Schrödinger equation so that multiple worlds never develop, adding new variables in addition to the wave function, or reinterpreting the wave function as a statement about our knowledge rather than a direct description of reality. All of these roads have been enthusiastically walked down.

We turn first to the possibility of altering the Schrödinger equation. This approach would seem to be squarely in the comfort zone of most physicists; almost before any successful theory has been established, theorists ask how they could play around with the underlying equations to make it even better. Schrödinger himself originally hoped that his equation would describe waves that naturally localized into blobs that behaved like particles when viewed from far away. Perhaps some modification of his equation could achieve that ambition, and even provide a natural resolution to the measurement problem without permitting multiple worlds.

This is harder than it sounds. If we try the most obvious thing, adding new terms like Ψ^2 to the equation, we tend to ruin important features of the theory, such as the total set of probabilities adding up to one. This kind of obstacle rarely deters physicists. Steven Weinberg, who developed the successful model that unified the electromagnetic and weak interactions in the Standard Model of particle physics, proposed a clever modification of the Schrödinger equation that manages to maintain the total probability over time. It comes at a cost, however; the simplest version of Weinberg's theory allows you to send signals faster than light between entangled particles, as opposed to the no-signaling theorem of ordinary quantum mechanics. This flaw can be patched, but then something even weirder occurs: not only are there still other branches of the wave function, but you can actually send signals between them, building what physicist Joe Polchinski dubbed an "Everett phone." Maybe that's a good thing, if you want to base your life choices on the outcome of a quantum measurement and then check in with your alternate selves to see which one turned out the best. But it doesn't seem to be the way that nature actually works. And it doesn't succeed in solving the measurement problem or getting rid of other worlds.

In retrospect this makes sense. Consider an electron in a pure spin-up state. That can equally well be expressed as an equal superposition of spin-left and spin-right, so that an observation along a horizontal magnetic field has a 50 percent chance of observing either outcome. But precisely because of that equality between the two options, it's hard to imagine how a deterministic equation could predict that we would see either one or the other (at least without the addition of new variables carrying additional information). Something would have to break the balance between spin-left and spin-right.

We therefore have to think a bit more dramatically. Rather than taking the Schrödinger equation and gently tinkering with it, we can bite the bullet and introduce a completely separate way for wave functions to evolve, one that squelches the appearance of multiple branches. Plenty of experimental evidence assures us that wave functions *usually* obey the Schrödinger equation, at least when we're not observing them. But maybe, rarely but crucially, they do something very different.

What might that different thing be? We seek to avoid the existential horror of multiple copies of the macroscopic world being described in a single wave function. So what if we imagined that wave functions undergo occasional *spontaneous collapse*, converting suddenly from being spread out over different possibilities (say, positions in space) to being relatively well localized around just one point? This is the key new feature of dynamical-collapse models, the most famous of which is *GRW theory*, after its inventors Giancarlo Ghirardi, Alberto Rimini, and Tullio Weber.

Envision an electron in free space, not bound to any atomic nucleus. According to the Schrödinger equation, the natural evolution of such a particle is for its wave function to spread out and become increasingly diffuse. To this picture, GRW adds a postulate that says at every moment there is some probability that the wave function will change radically and instantaneously. The peak of the new wave function is itself chosen from a probability distribution, the same one that we would have used to predict the position we would measure for the electron according to its original wave function. The new wave function is strongly concentrated around this central point, so that the particle is now essentially in one location as far as we macroscopic observers are concerned. Wave function collapses in GRW are real and random, not induced by measurements.

GRW theory is not some nebulous "interpretation" of quantum mechanics; it is a brand-new physical theory, with different dynamics. In fact, the theory postulates two new constants of nature: the width of the newly localized wave function, and the probability per second that the dynamical collapse will occur. Realistic values for these parameters are perhaps 10^{-5} centimeters for the width, and 10^{-16} for the probability of collapse per second. A typical electron therefore evolves for 10^{16} seconds before its wave function spontaneously collapses. That's about 300 million years. So in the 14-billion-year lifetime of the observable universe, most electrons (or other particles) localize only a handful of times.

That's a feature of the theory, not a bug. If you're going to go messing around with the Schrödinger equation, you had better do it in such a way as to not ruin all of the wonderful successes of conventional quantum mechanics. We do quantum experiments all the time with single particles or collections of a few particles. It would be disastrous if the wave functions of those particles kept spontaneously collapsing on us. If there is a truly random element in the evolution of quantum systems, it should be incredibly rare for individual particles.

Then how does such a mild alteration of the theory manage to get rid of macroscopic superpositions? Entanglement comes to the rescue, much as it did with decoherence in Many-Worlds.

Consider measuring the spin of an electron. As we pass it through a Stern-Gerlach magnet, the wave function of the electron evolves into a superposition of "deflected upward" and "deflected downward." We measure which way it went, for example, by detecting the deflected electron on a screen, which is hooked up to a dial with a pointer indicating Up or Down. An Everettian says that the pointer is a big macroscopic object that quickly becomes entangled with the environment, leading to decoherence and branching of the wave function. GRW can't appeal to such a process, but something related happens.

It's not that the original electron spontaneously collapses; we would have to wait for millions of years for that to become a likely event. But the pointer in the apparatus contains something like 10^{24} electrons, protons, and neutrons. All of these particles are entangled in an obvious way: they are in different positions depending on whether the pointer indicates Up or Down. Even though it's quite unlikely that any specific particle will undergo spontaneous collapse before we open the box, chances are extremely good that at least one of them will—that should happen roughly 10^8 times per second.

You might not be impressed, thinking that we wouldn't even notice a tiny subset of particles becoming localized in a macroscopic pointer. But the magic of entanglement means that if the wave function of just one particle is spontaneously localized, the rest of the particles with which that one is entangled will come along with it. If somehow the pointer did manage to avoid any of its particles localizing for a certain period of time, enough for it to evolve into a macroscopic superposition of Up and Down, that superposition would instantly collapse as soon as just one of the particles did localize. The overall wave function goes very rapidly from describing an apparatus pointing in a superposition of two answers to one that is definitively one or the other. GRW theory manages to make operational and objective the classical/quantum split that partisans of the Copenhagen approach are forced to invoke. Classical behavior is seen in objects that contain so many particles that it becomes likely that the overall wave function will undergo a series of rapid collapses.

GRW theory has obvious advantages and disadvantages. The primary advantage is that it's a well-posed, specific theory that addresses the measurement problem in a straightforward way. The multiple worlds of the Everett approach are eliminated by a series of truly unpredictable collapses. We are left with a world that maintains the successes of quantum theory in the microscopic realm, while exhibiting classical behavior macroscopically. It is a perfectly realist account that doesn't invoke any fuzzy notions about consciousness in its explanation of experimental outcomes. GRW can be thought of as Everettian quantum mechanics plus a random process that cuts off new branches of the wave function as they appear.

Moreover, it is experimentally testable. The two parameters governing the width of localized wave functions and the probability of collapse were not chosen arbitrarily; if their values were very different, they either wouldn't do the job (collapses would be too rare, or not sufficiently localized) or they would already have been ruled out by experiment. Imagine we have a fluid of atoms in an incredibly low-temperature state, so that every atom is moving very slowly if at all. A spontaneous collapse of the wave function of any electron in the fluid would give its atom a little jolt of energy, which physicists could detect as a slight increase in the temperature of the fluid. Experiments of this form are ongoing, with the ultimate goal of either confirming GRW, or ruling it out entirely.

These experiments are easier said than done, as the amount of energy we're talking about is very small indeed. Still, GRW is a great example to bring up when your friends complain that Many-Worlds, or different approaches to quantum mechanics more generally, aren't experimentally testable. You test theories in comparison to other theories, and these two are manifestly different in their empirical predictions.

Among GRW's disadvantages are the fact that, well, the new spontaneous-collapse rule is utterly ad hoc and out of step with everything else we know about physics. It seems suspicious that nature would not only choose to violate its usual law of motion at random intervals but do so in just such a way that we wouldn't yet have been able to experimentally detect it.

Another disadvantage, one that has prevented GRW and related theories from gaining traction among theoretical physicists, is that it's unclear how to construct a version of the theory that works not only for particles but also for fields. In modern physics, the fundamental building blocks of nature are fields, not particles. We see particles when we look closely enough at vibrating fields, simply because those fields obey the rules of quantum mechanics. Under some conditions, it's possible to think of the field description as useful but not mandatory, and imagine that fields are just ways of keeping track of many particles at once. But there are other circumstances (such as in the early universe, or inside protons and neutrons) where the field-ness is indispensable. And GRW, at least in the simple version presented here, gives us instructions for how wave functions collapse that refers specifically to the probability per particle. This isn't necessarily an insurmountable obstacle—taking simple models that don't quite work and generalizing them until they do is the theoretical physicist's stock-in-trade—but it's a sign that these approaches don't seem to fit naturally with how we currently think about the laws of nature.

GRW delineates the quantum/classical boundary by making spontaneous collapses very rare for individual particles, but very rapid for large collections. An alternative approach would be to make collapse occur whenever the system reached a certain threshold, like a rubber band breaking when it is stretched too far. A well-known example of an attempt along these lines was put forward by mathematical physicist Roger Penrose, best known for his work in general relativity. Penrose's theory uses gravity in a crucial way. He suggests that wave functions spontaneously collapse when they begin to describe macroscopic superpositions in which different components have appreciably different gravitational fields. The criterion of "appreciably different" here turns out to be difficult to specify precisely; single electrons would not collapse no matter how spread-out their wave functions were, while a pointer is large enough to cause collapse as soon as it started evolving into different states.

Most experts in quantum mechanics have not warmed to Penrose's theory, in part because they are skeptical that gravity should have anything to do with the fundamental formulation of quantum mechanics. Surely, they think, we can talk—and did, for most of the history of the subject—about quantum mechanics and wave-function collapse without considering gravity at all.

It's possible that a precise version of Penrose's criterion could be developed in which it is thought of as decoherence in disguise: the gravitational field of an object can be thought of as part of its environment, and if two different components of the wave function have different gravitational fields, they become effectively decohered. Gravity is an extremely weak force, and it will almost always be the case that ordinary electromagnetic interactions will cause decoherence long before gravity would. But the nice thing about gravity is that it's universal (everything has a gravitational field, not everything is electrically charged), so at least this would be a way to guarantee that the wave function would collapse for any macroscopic object. On the other hand, branching when decoherence occurs is already part of the Many-Worlds approach; all that this kind of spontaneous-collapse theory would say is "It's just like Everett, except that when new worlds are created, we erase them by hand." Who knows? That might be how nature actually works, but it's not a route that most working physicists are encouraged to pursue.

○ ○ ○

Since the very beginning of quantum mechanics, an obvious possibility to contemplate has been the idea that the wave function isn't the whole story, but that there are also other physical variables in addition to it. After all, physicists were very used to thinking in terms of probability distributions from their experience with statistical mechanics, as it had been developed in the nineteenth century. We don't specify the exact position and velocity of every atom in a box of gas, only their overall statistical properties. But in the classical view we take for granted that there is some exact position and velocity for each particle, even if we don't know it. Maybe quantum mechanics is like that—there are definite quantities associated with prospective observational outcomes, but we don't know what they are, and the wave function somehow captures part of the statistical reality without telling the whole story.

We know the wave function can't be exactly like a classical probability distribution. A true probability distribution assigns probabilities directly to outcomes, and the probability of any given event has to be a real number between zero and one (inclusive). A wave function, meanwhile, assigns an amplitude to every possible outcome, and amplitudes are complex numbers. They have both a real and an imaginary part, either one of which could be either positive or negative. When we square such amplitudes we obtain a probability distribution, but if we want to explain what is experimentally observed, we can't work directly with that distribution rather than keeping the wave function around. The fact that amplitudes can be negative allows for the interference that we see in the double-slit experiment, for example.

There's a simple way of addressing this problem: think of the wave function as a real, physically existing thing (not just a convenient summary of our incomplete knowledge), but *also* imagine that there are additional variables, perhaps representing the positions of particles. These extra quantities are conventionally called *hidden variables*, although some proponents of this approach don't like the label, as it's these variables that we actually observe when we make a measurement. We can just call them particles, since that's the case that is usually considered. The wave function then takes on the role of a *pilot wave*, guiding the particles as they move around. It's like particles are little floating barrels, and the wave function describes waves and currents in the water that push the barrels around. The wave function obeys the ordinary Schrödinger equation, while a new "guidance equation" governs how it influences the particles. The particles are guided to where the wave function is large, and away from where it is nearly zero.

The first such theory was presented by Louis de Broglie, at the 1927 Solvay Conference. Both Einstein and Schrödinger were thinking along similar lines at the time. But de Broglie's ideas were harshly criticized at Solvay, by Wolfgang Pauli in particular. From the records of the conference, it seems as if Pauli's criticisms were misplaced, and de Broglie actually answered them correctly. But he was sufficiently discouraged by the reception that de Broglie abandoned the idea.

In a famous book from 1932, *Mathematical Foundations of Quantum Mechanics*, John von Neumann proved a theorem about the difficulty of constructing hidden-variable theories. Von Neumann was one of the most brilliant mathematicians and physicists of the twentieth century, and his name carried enormous credibility among researchers in quantum mechanics. It became standard practice, whenever anyone would suggest that there might be a more definite way to formulate quantum theory than the vagueness inherent in the Copenhagen approach, for someone to invoke the name of von Neumann and the existence of his proof. That would squelch any budding discussion.

In fact what von Neumann had proven was something a bit less than most people assumed (often without reading his book, which wasn't translated into English until 1955). A good mathematical theorem establishes a result that follows from clearly stated assumptions. When we would like to invoke such a theorem to teach us something about the real world, however, we have to be very careful that the assumptions are actually true in reality. Von Neumann made assumptions that, in retrospect, we don't have to make if our task is to invent a theory that reproduces the predictions of quantum mechanics. He proved something, but what he proved was not "hidden-variable theories can't work." This was pointed out by mathematician and philosopher Grete Hermann, but her work was largely ignored.

Along came David Bohm, an interesting and complicated figure in the history of quantum mechanics. As a graduate student in the early 1940s, Bohm became interested in left-wing politics. He ended up working on the Manhattan Project, but he was forced to do his work in Berkeley, as he was denied the necessary security clearance to move to Los Alamos. After the war he became an assistant professor at Princeton, and published an influential textbook on quantum mechanics. In that book he adhered carefully to the received Copenhagen approach, but thinking through the issues made him start wondering about alternatives.

Bohm's interest in these questions was encouraged by one of the few figures who had the stature to stand up to Bohr and his colleagues: Einstein himself. The great man had read Bohm's book, and summoned the young professor to his office to talk about the foundations of quantum theory. Einstein explained his basic objections, that quantum mechanics couldn't be considered a complete view of reality, and encouraged Bohm to think more deeply about the question of hidden variables, which he proceeded to do.

All this took place while Bohm was under a cloud of political suspicion, at a time when association with Communism could ruin people's careers. In 1949, Bohm had testified before the House Un-American Activities Committee, where he refused to implicate any of his former colleagues. In 1950 he was arrested in his office at Princeton for contempt of Congress. Though he was eventually cleared of all charges, the president of the university forbade him from setting foot on campus, and put pressure on the physics department to not renew his contract. In 1951, with support from Einstein and Oppenheimer, Bohm was eventually able to find a job at the University of São Paulo, and left for Brazil. That's why the first seminar at Princeton to explain Bohm's ideas had to be given by someone else.

° ° °

None of this drama prevented Bohm from thinking productively about quantum mechanics. Encouraged by Einstein, he developed a theory that was similar to that of de Broglie, in which particles were guided by a "quantum potential" constructed from the wave function. Today this approach is often known as the *de Broglie–Bohm theory*, or simply *Bohmian mechanics*. Bohm's presentation of the theory was a bit more fleshed out than de Broglie's, especially when it came to describing the measurement process.

Even today you will sometimes hear professional physicists say that it's impossible to construct a hidden-variable theory that reproduces the predictions of quantum mechanics, "because of Bell's theorem." But that's exactly what Bohm did, at least for the case of non-relativistic particles. John Bell, in fact, was one of the few physicists who was extremely impressed by Bohm's work, and he was inspired to develop his theorem precisely to understand how to reconcile the existence of Bohmian mechanics with the purported no-hidden-variables theorem of von Neumann.

What Bell's theorem actually proves is the impossibility of reproducing quantum mechanics via a *local* hidden-variables theory. Such a theory is what Einstein had long been hoping for: a model that would attach independent reality to physical quantities associated with specific locations in space, with effects between them propagating at or below the speed of light. Bohmian mechanics is perfectly deterministic, but it is resolutely nonlocal. Separated particles can affect each other instantaneously.

Bohmian mechanics posits both a set of particles with definite (but unknown to us, until they are observed) positions, and a separate wave function. The wave function evolves exactly according to the Schrödinger equation—it doesn't even seem to recognize that the particles are there, and is unaffected by what they are doing. The particles, meanwhile, are pushed around according to a guidance equation that depends on the wave function. However, the way in which any one particle is guided depends not just on the wave function but also on the positions of *all the other particles* that may be in the system. That's the nonlocality; the motion of a particle here can depend, in principle, on the positions of other particles arbitrarily far away. As Bell himself later put it, in Bohmian mechanics "the Einstein-Podolsky-Rosen paradox is resolved in the way which Einstein would have liked least."

This nonlocality plays a crucial role in understanding how Bohmian mechanics reproduces the predictions of ordinary quantum mechanics. Consider the double-slit experiment, which illustrates so vividly how quantum phenomena are simultaneously wave-like (we see interference patterns) and particle-like (we see dots on the detector screen, and interference goes away when we detect which slit the particles go through). In Bohmian mechanics this ambiguity is not mysterious at all: there are both particles and waves. The particles are what we observe; the wave function affects their motion, but we have no way of measuring it directly.

According to Bohm, the wave function evolves through both slits just as it would in Everettian quantum mechanics. In particular, there will be interference effects where the wave function adds or cancels once it reaches the screen. But we don't see the wave function at the screen; we see individual particles hitting it. The particles are pushed around by the wave function, so that they are more likely to hit the screen where the wave function is large, and less likely to do so where it is small.

The Born rule tells us that the probability of observing a particle at a given location is given by the wave function squared. On the surface, this seems hard to reconcile with the idea that particle positions are completely independent variables that we can specify as we like. And Bohmian mechanics is perfectly deterministic—there aren't any truly random events, as there are with the spontaneous collapses of GRW theory. So where does the Born rule come from?

The answer is that, while in principle particle positions could be anywhere at all, in practice there is a natural distribution for them to have. Imagine that we have a wave function and some fixed number of particles. To recover the Born rule, all we have to do is start with a Born rule–like distribution of those particles. That is, we have to distribute the positions of our particles so that the distribution looks like it was chosen randomly with probability given by the wave function squared. More particles where the amplitude is large, fewer particles where it is small.

Such an "equilibrium" distribution has the nice feature that the Born rule remains valid as time passes and the system evolves. If we start our particles in a probability distribution that matches what we expect from ordinary quantum mechanics, it will continue to match that expectation going forward. It is believed by many Bohmians that a non-equilibrium initial distribution will evolve toward equilibrium, just as a gas of classical particles in a box evolves toward an equilibrium thermal state; but the status of this idea is not yet settled. The resulting probabilities are, of course, about our knowledge of the system rather than about objective frequencies; if somehow we knew exactly what the particle positions were, rather than just their distribution, we could predict experimental outcomes exactly without any need for probabilities at all.

This puts Bohmian mechanics in an interesting position as an alternative formulation of quantum mechanics. GRW theory matches traditional quantum expectations usually, but also makes definite predictions for new phenomena that can be tested. Like GRW, Bohmian mechanics is unambiguously a different physical theory, not simply an "interpretation." It doesn't *have* to obey the Born rule if for some reason our particle positions are not in an equilibrium distribution. But it will obey the rule if they are. And if that's the case, the predictions of Bohmian mechanics are strictly indistinguishable from those of ordinary quantum theory. In particular, we will see more particles hit the screen where the wave function is large, and fewer where it is small.

We still have the question of what happens when we look to see which slit the particle has gone through. Wave functions don't collapse in Bohmian mechanics; as with Everett, they always obey the Schrödinger equation. So how are we supposed to explain the disappearance of the interference pattern in the double-slit experiment?

The answer is "the same way we do in Many-Worlds." While the wave function doesn't collapse, it does evolve. In particular, we should consider the wave function for the detection apparatus as well as for the electrons going through the slits; the Bohmian world is completely quantum, not stooping to an artificial split between classical and quantum realms. As we know from thinking about decoherence, the wave function for the detector will become entangled with that of an electron passing through the slit, and a kind of "branching" will occur. The difference is that the variables describing the apparatus (which aren't there in Many-Worlds) will be at locations corresponding to one of these branches, and not the other. For all intents and purposes, it's just like the wave function has collapsed; or, if you prefer, it's just like decoherence has branched the wave function, but instead of assigning reality to each of the branches, the particles of which we are made are only located on one particular branch.

You won't be surprised to hear that many Everettians are dubious about this kind of story. If the wave function of the universe simply obeys the Schrödinger equation, it will undergo decoherence and branching. And you've already admitted that the wave function is part of reality. The particle positions, for that matter, have absolutely no influence on how the wave function evolves. All they do, arguably, is point to a particular branch of the wave function and say, "This is the real one." Some Everettians have therefore claimed that Bohmian mechanics isn't really any different from Everett, it just includes some superfluous extra variables that serve no purpose but to assuage some anxieties about splitting into multiple copies of ourselves. As Deutsch has put it, "Pilot-wave theories are parallel-universe theories in a state of chronic denial."

We won't adjudicate this dispute right here. What's clear is that Bohmian mechanics is an explicit construction that does what many physicists thought was impossible: to construct a precise, deterministic theory that reproduces all of the predictions of textbook quantum mechanics, without requiring any mysterious incantations about the measurement process or a distinction between quantum and classical realms. The price we pay is explicit nonlocality in the dynamics.

○ ○ ○

Bohm was hopeful that his new theory would be widely appreciated by physicists. This was not to be. In the emotionally charged language that so often accompanies discussions of quantum foundations, Heisenberg called Bohm's theory "a superfluous ideological superstructure," while Pauli referred to it as "artificial metaphysics." We've already heard the judgment of Oppenheimer, who had previously been Bohm's mentor and supporter. Einstein seems to have appreciated Bohm's effort, but thought the final construction was artificial and unconvincing. Unlike de Broglie, however, Bohm didn't bow to the pressure, and continued to develop and advocate for his theory. Indeed, his advocacy inspired de Broglie himself, who was still around and active (he died in 1987). In his later years de Broglie returned to hidden-variable theories, developing and elaborating his original model.

Even apart from the presence of explicit nonlocality and the accusation that the theory is just Many-Worlds in denial, there are other significant problems inherent in Bohmian mechanics, especially from the perspective of a modern fundamental physicist. The list of ingredients in the theory is undoubtedly more complicated than in Everett, and Hilbert space, the set of all possible wave functions, is as big as ever. The possibility of many worlds is not avoided by erasing the worlds (as in GRW), but simply by denying that they're real. The way Bohmian dynamics works is far from elegant. Long after classical mechanics was superseded, physicists still intuitively cling to something like Newton's third law: if one thing pushes on another, the second thing pushes back. It therefore seems strange that we have particles that are pushed around by a wave function, while the wave function is completely unaffected by the particles. Of course, quantum mechanics inevitably forces us to confront strange things, so perhaps this consideration should not be paramount.

More important, the original formulations of de Broglie and Bohm both rely heavily on the idea that what really exists are "particles." Just as with GRW, this creates a problem when we try to understand the best models of the world that we actually have, which are quantum field theories. People have proposed ways of "Bohmizing" quantum field theory, and there have been some successes—physicists can be extremely clever when they want to be. But the results feel forced rather than natural. It doesn't mean they are necessarily wrong, but it's a strike against Bohmian theories when compared to Many-Worlds, where including fields or quantum gravity is straightforward.

In our discussion of Bohmian mechanics we referred to the positions of the particles, but not to their momenta. This hearkens back to the days of Newton, who thought of particles as having a position at every moment in time, and velocity (and momentum) as derived from that trajectory, by calculating its rate of change. More modern formulations of classical mechanics (well, since 1833) treat position and momentum on an equal footing. Once we go to quantum mechanics, this perspective is reflected in the Heisenberg uncertainty principle, in which position and momentum appear in exactly the same way. Bohmian mechanics undoes this move, treating position as primary, and momentum as something that derived from it. But it turns out that you can't measure it exactly, due to unavoidable effects of the wave function on the particle positions over time. So at the end of the day, the uncertainty principle remains true in Bohmian mechanics as a practical fact of life, but it doesn't have the automatic naturalness of theories in which the wave function is the only real entity.

There is a more general principle at work here. The simplicity of Many-Worlds also makes it extremely flexible. The Schrödinger equation takes the wave function and figures out how fast it will evolve by applying the Hamiltonian, which measures the different amounts of energy in different components of the quantum state. You give me a Hamiltonian, and I can instantly understand the Everettian version of its corresponding quantum theory. Particles, spins, fields, superstrings, doesn't matter. Many-Worlds is plug-and-play.

Other approaches require a good deal more work than that, and it's far from clear that the work is even doable. You have to specify not only a Hamiltonian but also a particular way in which wave functions spontaneously collapse, or a particular new set of hidden variables to keep track of. That's easier said than done. The problem becomes even more pronounced when we move from quantum field theory to quantum gravity (which, remember, was one of Everett's initial motivations). In quantum gravity the very notion of "a location in space" becomes problematic, as different branches of the wave function will have different spacetime geometries. For Many-Worlds that's no problem; for alternatives it's close to a disaster.

When Bohm and Everett were inventing their alternatives to Copenhagen in the 1950s, or Bell was proving his theorems in the 1960s, work on foundations of quantum mechanics was shunned within the physics community. That began to change somewhat with the advent of decoherence theory and quantum information in the 1970s and '80s;

GRW theory was proposed in 1985. While this subfield is still looked upon with suspicion by a large majority of physicists (for one thing, it tends to attract philosophers), an enormous amount of interesting and important work has been accomplished since the 1990s, much of it wide out in the open. However, it's also safe to say that much contemporary work on quantum foundations still takes place in a context of qubits or non-relativistic particles. Once we graduate to quantum fields and quantum gravity, some things we could previously take for granted are no longer available. Just as it is time for physics as a field to take quantum foundations seriously, it's time for quantum foundations to take field theory and gravity seriously.

° ° °

In contemplating ways to eliminate the many worlds implied by a bare-bones version of the underlying quantum formalism, we have explored chopping off the worlds by a random event (GRW) or reaching some kind of threshold (Penrose) or picking out particular worlds as real by adding additional variables (de Broglie–Bohm). What's left?

The problem is that the appearance of multiple branches of the wave function is automatic once we believe in wave functions and the Schrödinger equation. So the alternatives we have considered thus far either eliminate those branches or posit something that picks out one of them as special.

A third way suggests itself: deny the reality of the wave function entirely.

By this we don't mean to deny the central importance of wave functions in quantum mechanics. Rather, we can use wave functions, but we might not claim that they represent part of reality. They might simply characterize our knowledge; in particular, the incomplete knowledge we have about the outcome of future quantum measurements. This is known as the "epistemic" approach to quantum mechanics, as it thinks of wave functions as capturing something about what we know, as opposed to "ontological" approaches that treat the wave function as describing objective reality. Since wave functions are usually denoted by the Greek letter Ψ (Psi), advocates of epistemic approaches to quantum mechanics sometimes tease Everettians and other wave-function-realists by calling them "Psi-ontologists."

We've already noted that an epistemic strategy cannot work in the most naïve and straightforward way. The wave function is not a probability distribution; real probability distributions are never negative, so they can't lead to interference phenomena such as we observe in the double-slit experiment. Rather than giving up, however, we can try to be a bit more sophisticated in how we think about the relationship between the wave function and the real world. We can imagine building up a formalism that allows us to use wave functions to calculate the probabilities associated with experimental outcomes, while not attaching any underlying reality to them. This is the task taken up by epistemic approaches.

There have been many attempts to interpret the wave function epistemically, just as there are competing collapse models or hidden-variable theories. One of the most prominent is Quantum Bayesianism, developed by Christopher Fuchs, Rüdiger Schack, Carlton Caves, N. David Mermin, and others. These days the label is typically shortened to QBism and pronounced "cubism." (One must admit it's a charming name.)

Bayesian inference suggests that we all carry around with us a set of credences for various propositions to be true or false, and update those credences when new information comes in. All versions of quantum mechanics (and indeed all scientific theories) use Bayes's theorem in some version or another, and in many approaches to understanding quantum probability it plays a crucial role. QBism is distinguished by making our quantum credences *personal*, rather than universal. According to QBism, the wave function of an electron isn't a once-and-for-all thing that everyone could, in principle, agree on. Rather, everyone has their own idea of what the electron's wave function is, and uses that idea to make predictions about observational outcomes. If we do many experiments and talk to one another about what we've observed, QBists claim, we will come to a degree of consensus about what the various wave functions are. But they are fundamentally measures of our personal belief, not objective features of the world. When we see an electron deflected upward in a Stern-Gerlach magnetic field, the world doesn't change, but we've learned something new about it.

There is one immediate and undeniable advantage of such a philosophy: if the wave function isn't a physical thing, there's no need to fret about it "collapsing," even if that collapse is purportedly nonlocal. If Alice and Bob possess two particles that are entangled with each other and Alice makes a measurement, according to the ordinary rules of quantum mechanics the state of Bob's particle changes instantaneously. QBism reassures us that we needn't worry about that, as there is no such thing as "the state of Bob's particle." What changed was the wave function that Alice carries around with her to make predictions: it was updated using a suitably quantum version of Bayes's theorem. Bob's wave function didn't change at all. QBism arranges the rules of the game so that when Bob does get around to measuring his particle, the outcome will agree with the prediction we would make on the basis of Alice's measurement outcome. But there is no need along the way to imagine that any physical quantity changed over at Bob's location. All that changes are different people's states of knowledge, which after all are localized in their heads, not spread through all space.

Thinking about quantum mechanics in QBist terms has led to interesting developments in the mathematics of probability, and offers insight into quantum information theory. Most physicists, however, will still want to know: What is reality supposed to be in this view? (Abraham Pais recalled that Einstein once asked him whether he "really believed that the moon exists only when I look at it.")

The answer is not clear. Imagine that we send an electron through a

Stern-Gerlach magnet, but we choose not to look at whether it's deflected up or down. For an Everettian, it is nevertheless the case that decoherence and branching has occurred, and there is a fact of the matter about which branch any particular copy of ourselves is on. The QBist says something very different: there is no such thing as whether the spin was deflected up or down. All we have is our degrees of belief about what we will see when we eventually decide to look. There is no spoon, as Neo learned in *The Matrix*. Fretting about the "reality" of what's going on before we look, in this view, is a mistake that leads to all sorts of confusion.

QBists, for the most part, don't talk about what the world really is. Or at least, as an ongoing research program, QBists have chosen not to dwell too much on the questions concerning the nature of reality about which the rest of us care so much. The fundamental ingredients of the theory are a set of *agents*, who have *beliefs*, and accumulate *experiences*. Quantum mechanics, in this view, is a way for agents to organize their beliefs and update them in the light of new experiences. The idea of an agent is absolutely central; this is in stark contrast to the other formulations of quantum theory that we've been discussing, according to which observers are just physical systems like anything else.

Sometimes QBists will talk about reality as something that comes into existence as we make observations. Mermin has written, "There is indeed a common external world in addition to the many distinct individual personal external worlds. But that common world must be understood at the foundational level to be a mutual construction that all of us have put together from our distinct private experiences, using our most powerful human invention: language." The idea is not that there is no reality, but that reality is more than can be captured by any seemingly objective third-person perspective. Fuchs has dubbed this view *Participatory Realism*: reality is the emerging totality of what different observers experience.

QBism is relatively young as approaches to quantum foundations go, and there is much development yet to be done. It's possible that it will run into insurmountable roadblocks, and interest in the ideas will fizzle out. It's also possible that the insights of QBism can be interpreted as a sometimes-useful way of talking about the experiences of observers within some other, straightforwardly realist, version of quantum mechanics. And finally, it might be that QBism or something close to it represents a true, revolutionary way of thinking about the world, one that puts agents like you and me at the center of our best description of reality.

Personally, as someone who is quite comfortable with Many-Worlds (while recognizing that we still have open questions), this all seems to me like an incredible amount of effort devoted to solving problems that aren't really there. QBists, to be fair, feel a similar level of exasperation with Everett: Mermin has said that "QBism regards [branching into many simultaneously existing worlds] as the *reductio ad absurdum* of reifying the quantum state." That's quantum mechanics for you, where one person's absurdity is another person's answer to all of life's questions.

◦ ◦ ◦

The foundations-of-physics community, which is full of smart people who have thought long and hard about these issues, has not reached a consensus on the best approach to quantum mechanics. One reason is that people come to the problem from different backgrounds, and therefore with different concerns foremost in their minds. Researchers in fundamental physics—particle theory, general relativity, cosmology, quantum gravity—tend to favor the Everett approach, if they deign to take a position on quantum foundations at all. That's because Many-Worlds is extremely robust to the underlying physical stuff it is describing. You give me a set of particles and fields and what have you, and rules for how they interact, and it's straightforward to fit those elements into an Everettian picture. Other approaches tend to be more persnickety, demanding that we start from scratch to figure out what the theory actually says in each new instance. If you're someone who admits that we don't really know what the underlying theory of particles and fields and spacetime really is, that sounds exhausting, whereas Many-Worlds is a natural easy resting place. As David Wallace has put it, "The Everett interpretation (insofar as it is philosophically acceptable) is the only interpretative strategy currently suited to make sense of quantum physics as we find it."

But there is another reason, more based in personal style. Essentially everyone agrees that simple, elegant ideas are to be sought after as we search for scientific explanations. Being simple and elegant doesn't mean an idea is correct—that's for the data to decide—but when there are multiple ideas vying for supremacy and we don't yet have enough data to choose among them, it's natural to give a bit more credence to the simplest and most elegant ones.

The question is, who decides what's simple and elegant? There are different senses of these terms. Everettian quantum mechanics is absolutely simple and elegant from a certain point of view. A smoothly evolving wave function, that's all. But the result of these elegant postulates—a proliferating tree of multiple universes—is arguably not very simple at all.

Bohmian mechanics, on the other hand, is constructed in a kind of haphazard way. There are both particles and wave functions, and they interact through a nonlocal guidance equation that seems far from elegant. Including both particles and wave functions as fundamental ingredients is, however, a natural strategy to contemplate, once we have been confronted with the basic experimental demands of quantum mechanics. Matter acts sometimes like waves and sometimes like particles, so we invoke both waves and particles. GRW theory, meanwhile, adds a weird ad hoc stochastic modification to the Schrödinger equation. But it's arguably the simplest, most brute-force way to physically implement the fact that wave functions appear to collapse.

There is a useful contrast to be drawn between the simplicity of a physical theory and the simplicity with which that theory maps onto reality as we observe it. In terms of basic ingredients, Many-Worlds is unquestionably as simple as it gets. But the distance between what the theory itself says (wave functions, Schrödinger equation) and what we observe in the world (particles, fields, spacetime, people, chairs, stars, planets) seems enormous. Other approaches might be more baroque in their underlying principles, but it's relatively clear how they account for what we see.

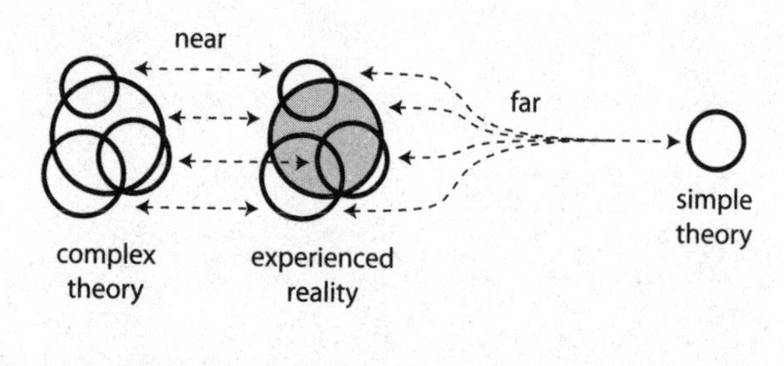

Both underlying simplicity and closeness to the phenomena are virtues in their own rights, but it's hard to know how to balance them against each other. This is where personal style comes in. All of the approaches to quantum mechanics that we've considered face looming challenges as we contemplate developing them into rock-solid foundations for an understanding of the physical world. So each of us has to make a personal judgment about which of these problems will eventually be solved, and which will prove fatal for the various approaches. That's okay; indeed, it's crucial that different people come down differently on these judgments about how to move forward. That gives us the best chance to keep multiple ideas alive, maximizing the probability that we'll eventually get things right.

Many-Worlds offers a perspective on quantum mechanics that is not only simple and elegant at its core but seems ready-made for adapting to the ongoing quest to understand quantum field theory and the nature of spacetime. That's enough to convince me that I should learn to live with the annoyance of other copies of me being produced all the time. But if it turns out that an alternative approach answers our deepest questions more effectively, I'll happily change my mind.

The Human Side

Living and Thinking in a Quantum Universe

In the course of a long life, each of us will occasionally encounter a difficult decision we must make. Stay single or get married? Go for a run or have another doughnut? Go to grad school or enter the real world?

Wouldn't it be nice to be able to choose both sides, rather than picking one? Quantum mechanics suggests a strategy: whenever you have a decision to make, you can do so by consulting a quantum random-number generator. Indeed, there is an app available for iPhones called Universe Splitter that can be used for this very purpose. (As Dave Barry says, I swear I am not making this up.)

Let's say you have a choice to make: "Should I get pepperoni or sausage on my pizza?" (And let's say you have too much restraint to give the obvious answer of asking for both on the same pizza.) You can fire up Universe Splitter, where you will see two text boxes, into which you can type "pepperoni" and "sausage." Then hit the button, and your phone will send a signal through the internet to a laboratory in Switzerland, where a photon is sent toward a beam splitter (essentially a partially silvered mirror that reflects some photons and lets others through).

According to the Schrödinger equation, the beam splitter turns the photon's wave function into two components going left and right, each of which heads toward a different detector. When either detector notices a photon, it produces a readout that becomes entangled with the environment, quickly leading to decoherence and branching the wave function in two. The copy of you in the branch where the photon went left sees their phone flash with the message "pepperoni," and in the one where it went right, they see "sausage." If each one actually follows up with your plan to do what your phone advises, there will be one world in which a version of you orders pepperoni, and another in which a version of you orders sausage. Sadly, the two persons have no way of communicating with each other to share tasting notes afterward.

Even for the most battle-hardened quantum physicist, one must admit that this *sounds* ludicrous. But it's the most straightforward reading of our best understanding of quantum mechanics.

The question naturally arises: What should we do about it? If the real world is truly this radically different from the world of our everyday experience, does this have any implications for how we live our lives?

Largely—no. To each individual on some branch of the wave function, life goes on just as if they lived in a single world with truly stochastic quantum events. But the issues are worth exploring.

∘ ∘ ∘

You are welcome to offload your hard decisions to a quantum random-number generator, thereby ensuring that there is at least one branch of the wave function in which the best alternative was chosen. But let's say we choose not to. Should the branching of our current selves into multiple future selves affect the choices we make? In the textbook view, there is a probability that one or another outcome happens when we observe a quantum system, while in Many-Worlds all outcomes happen, weighted by the amplitude squared of the wave function. Does the existence of all those extra worlds have implications for how we should act, personally or ethically?

It's not hard to imagine that it might, but upon careful consideration it turns out to matter much less than you might guess. Consider the infamous quantum suicide experiment, or the related idea of quantum immortality. It's an idea that has been considered ever since Many-Worlds came on the scene—reportedly Hugh Everett himself believed a version of quantum immortality—but has been popularized by physicist Max Tegmark.

Here's the setup: we imagine a deadly device that is triggered by a quantum measurement, such as sending a query to the Universe Splitter app. Imagine that the quantum measurement has a 50 percent chance of triggering a gun that shoots a bullet into my head at close range, and a 50 percent chance of doing nothing. According to Many-Worlds, that implies the existence of two branches of the wave function, one of which contains a living version of me, the other of which contains a dead version.

Assume for purposes of the thought experiment we believe that life itself is a purely physical phenomenon, so we can set aside considerations of life after death. From my perspective, the branch on which the gun fired isn't one that any version of me ever gets to experience—my descendant in that world is dead. But my descendant continues on, unharmed, on the branch where the gun didn't fire. In some sense, then, "I" will live forever, even if I repeat this macabre procedure over and over again. One might go so far as to argue that I shouldn't object to actually going through this experiment (putting aside the rest of the world's feelings about me, I suppose)—in the branches where the gun fired "I" don't really exist, while in the single branch where it failed to fire time after time I'm perfectly healthy. (Tegmark's original point was less grandiose: he simply noted that an experimenter who survived a large number of trials would have good reason to accept the Everett picture.) This conclusion stands in stark contrast to a conventional stochastic formulation of quantum mechanics, where there is only one world, and I would have an increasingly tiny chance of being alive within it.

I do *not* recommend that you try such an experiment at home. In fact, the logic behind not caring about those branches in which you are killed is more than a little wonky.

Consider life in an old-fashioned, classical, single-universe picture. If you thought you lived in such a universe, would you mind if someone sneaked up behind you and shot you in the head so that you died instantly? (Again, setting aside the possibility that other people might be upset.) Most of us would not be in favor of that happening. But by the logic above, you really shouldn't "mind"—after all, once you're dead, there's no "you" to be upset about what happened.

The point being missed by this analysis is that we are upset *now*—while we are still very much alive and feeling—by the prospect of being dead in the future, especially if that future comes sooner rather than later. And that's a valid perspective; much of how we think about our current lives depends on a projection into the rest of our existence. Cutting that existence off is something we are perfectly allowed to object to, even if we won't be around to be bothered by it once it happens. And given that, quantum suicide turns out to be just as bleak and unpalatable as our immediate intuition might suggest. It's okay for me to yearn for a happy and long life for all the future versions of me that will end up in various branches of the wave function, as much as it would be valid for me to hope for a long life if I thought there was just a single world.

This goes back to something we discussed in Chapter Seven: the importance of treating individuals on different branches of the wave function as distinct persons, even if they descended from the same individual in the past. There is an important asymmetry between how we think about "our future" versus "our past" in Many-Worlds, which ultimately can be attributed to the low-entropy condition of our early universe. Any one individual can trace their lives backward in a unique person, but going forward in time we will branch into multiple people. There is not one future self that is picked out as "really you," and it's equally true that there is no one person constituted by all of those future individuals. They are separate, as much as identical twins are distinct people, despite descending from a single zygote.

We might care about what happens to the versions of ourselves who live on other branches, but it's not sensible to think of them as "us." Imagine that you're just about to perform a vertical-spin measurement on an electron you have prepared in an equal superposition of spin-up and spin-down. A random philanthropist enters your lab and offers you the following bargain: if the spin is up, they will give you a million dollars; if the spin is down, you give them one dollar. You would be wise to take the deal; for all intents and purposes, it's as if you are being offered a bet with equal chances of winning a million dollars or losing just one dollar, even if one of your future selves will certainly be out a dollar.

But now imagine that you were a little quicker in your experimental setup, and you observed a spin-down outcome just before the philanthropist busts in. It turns out that they are a pushy deal-maker, and they explain that the version of you on the other branch is being given a million dollars, but you now have to give them one dollar in this branch.

There's no reason for you to be happy about this (or to give up the dollar), even though the version of you on the other branch might be happy about it. You are not them, and they are not part of you. Post-branching, you're two different people. Neither your experiences nor your rewards should be thought of as being shared by various copies of you on different branches. Don't play quantum Russian roulette, and don't accept losing bargains from pushy philanthropists.

○ ○ ○

That may be a reasonable policy when it comes to your own well-being, but what about that of others? How does knowing about the existence of other worlds affect our notions of moral or ethical behavior?

The right way to think about morality is itself a controversial subject, even in single-world versions of reality, but it's instructive to consider two broad categories of moral theory: *deontology* and *consequentialism*. Deontologists hold that moral behavior is a matter of obeying the right rules; actions are inherently right or wrong, whatever their consequences might turn out to be. Consequentialists, unsurprisingly, have the alternative view: we should work to maximize the beneficent consequences of our actions. Utilitarians, who advocate maximizing some measure of overall well-being, are paradigmatic consequentialists. There are other options, but these illustrate the basic point.

Deontology would seem to be unaffected by the possible presence of other worlds. If the whole point of your theory is that actions are intrinsically right or wrong, regardless of what outcomes they lead to, the existence of more worlds in which those outcomes can occur doesn't really matter. A typical deontological rule is Kant's categorical imperative: "Act only according to that maxim whereby you can, at the same time, will that it should become a universal law." It seems like it would be safe here to replace "a universal law" by "a law holding in all branches of the wave function," without altering any substantive judgment about what kind of actions might qualify.

Consequentialism is another matter entirely. Imagine that you are a no-nonsense utilitarian, who believes there is a quantity called *utility* that measures the amount of well-being associated with conscious creatures, and that this quantity can be added among all creatures to obtain a total utility, and that the morally right course of action is the one that maximizes this total utility. Imagine further that you judge the total utility in the entire universe to be some positive number. (If you didn't, you'd be in favor of trying somehow to destroy the universe, which makes for a good supervillain origin story but not for good neighbors.)

It would follow that, if the universe has positive utility and our goal is to maximize utility, creating a new copy of the whole universe would be one of the most morally valorous actions you could possibly take. The right thing to do would then be to branch the wave function of the universe as often as possible. We could imagine building a quantum utility maximizing device (QUMaD), perhaps an apparatus that continually bounces electrons through a device that measures first their vertical spin, then their horizontal spin. Every time an electron undergoes either measurement, the universe branches in two, doubling the total utility of all universes. Having built QUMaD and turned it on, you would be the most moral person ever to live!

Something about this smells fishy, however. Turning on QUMaD has no impact whatsoever on the lives of people in this universe or any other. They don't even know the machine exists. Are we really sure it has such a morally praiseworthy effect?

Happily there are a couple of ways out of this puzzle. One is to deny the assumptions: maybe this kind of no-nonsense utilitarianism isn't the best moral theory. There is a long and honorable tradition of people inventing things that would nominally increase the utility of the universe, but don't resemble our moral intuitions whatsoever. (Robert Nozick imagined a "utility monster," a hypothetical being that was so good at experiencing pleasure that the most moral thing anyone could do would be to keep the monster as happy as possible, no matter who else might suffer thereby.) QUMaD is just another example along these lines. The simple idea of adding up utilities among different people doesn't always lead to the results we might initially have imagined.

But there's another solution, one that comports more directly with the Many-Worlds philosophy. When we talked about deriving the Born rule, we discussed how to apportion credences in conditions of self-locating uncertainty: you know the wave function of the universe, but you don't know which branch you are on. The answer was that your credences should be proportional to the weight of the branch—the corresponding amplitude, squared. This "weight" is a crucially important aspect of how we think about worlds in an Everettian picture. It's not just probability that goes that way; conservation of energy also only works if we multiply the energy of each branch by its associated weight.

It makes sense, then, that we should do the same with utility. If we have a universe with some given total utility, and we measure a spin to branch it in two, the post-branching utility should be the sum of the weights of each branch times their utilities. Then, in the likely event that our spin measurement didn't affect anyone's utility in a substantial way, the total utility is completely unchanged by our measurement. That's just what our intuition might expect. It's also what we would directly conclude from the decision-theoretic approach to probability we mentioned in Chapter Six. From this perspective, Many-Worlds shouldn't change our ideas about moral action in any noticeable way.

It's nevertheless possible to cook up a system in which the difference between Many-Worlds and collapse theories really would be morally relevant. Imagine that some quantum experiment will lead to equally likely outcomes A or B, with A being extremely good and B being just a little bit good, and that these effects apply to everyone in the world with equal measure. In a single-world view, a utilitarian (or any commonsensical person, really) would be in favor of running the experiment, since either the vast good of A or the minor good of B would raise the net utility of the world. But imagine that your ethical code is entirely devoted to equality: you don't care what happens, as long as it happens to everyone equally. On the collapse theory, you don't know which outcome will happen, but either one maintains equality, so it's still a good idea to run the experiment. But in Many-Worlds, people in one branch will experience A while those on the other branch will experience B. Even if the branches can't communicate or otherwise interact, this could conceivably offend your moral sensibilities, so you'd be against doing the experiment at all. Personally I don't think that inequality between people who literally live in different worlds should matter that much to us, but the logical possibility is there.

Excluding such artificial constructions, Many-Worlds doesn't seem to have many moral implications. The picture of branching as "creating" an entirely new copy of the universe is a vivid one, but not quite right. It's better to think of it as dividing the existing universe into almost-identical slices, each one of which has a smaller weight than the original. If we follow that picture carefully, we conclude that it's correct to think about our future exactly as if we lived in a single stochastic universe that obeyed the Born rule. As counterintuitive as Many-Worlds might seem, at the end of the day it doesn't really change how we should go through our lives.

○ ○ ○

So far we've treated branching of the wave function as something that happens independently of ourselves, so that we simply have to go along for the ride. It's worth asking whether that's the proper perspective. Whenever I make a decision, are different worlds created where I chose different things? Are there realities out there corresponding to every series of alternative choices I could have made, universes that actualize all the possibilities of my life?

The idea of "making a decision" isn't something inscribed in the fundamental laws of physics. It's one of those useful, approximate, emergent notions that we find convenient to invoke when describing human-scale phenomena. What you and I label "making a decision" is a set of neurochemical processes happening in our brain. It's perfectly okay to talk about making decisions, but it's not something over and above ordinary material stuff obeying the laws of physics.

So the question is, do the physical processes going on in your brain when you make a decision *cause* the wave function of the universe to branch, with different decisions being made in each branch? If I'm playing poker and lose all my chips after making an ill-timed bluff, can I take solace in the idea that there is another branch where I played more conservatively?

No, you do not cause the wave function to branch by making a decision. In large part that's just due to what we mean (or ought to mean) by something "causing" something else. Branching is the result of a microscopic process amplified to macroscopic scales: a system in a quantum superposition becomes entangled with a larger system, which then becomes entangled with the environment, leading to decoherence. A decision, on the other hand, is a purely macroscopic phenomenon. There are no decisions being made by the electrons and atoms inside your brain; they're just obeying the laws of physics.

Decisions and choices and their consequences are useful concepts when we are talking about things at the macroscopic, human-size level. It's perfectly okay to think of choices as really existing and having influences, as long as we confine such talk to the regime in which they apply. We can choose, in other words, to talk about a person as a bunch of particles obeying Schrödinger's equation, or we can equally well talk about them as an agent with volition who makes decisions that affect the world. But we can't use both descriptions at once. Your decisions don't cause the wave function to branch, because "the wave function branching" is a relevant concept at the level of fundamental physics, and "your decisions" is a relevant concept at the everyday macroscopic level of people.

So there is no sense in which your decisions cause branching. But we can still ask whether there are other branches where you made different decisions. And indeed there might be, but the right way to think about the causality is "some microscopic process happened that caused branching, and on different branches you ended up making different decisions," rather than "you made a decision, which caused the wave function of the universe to branch." For the most part, however, when you do make a decision—even one that seems like a close call at the time—almost all of the weight will be concentrated on a single branch, not spread equally over many alternatives.

The neurons in our brains are cells consisting of a central body and a number of appendages. Most of those appendages are dendrites, which take in signals from surrounding neurons, but one of them is the axon, a longer fiber down which outgoing signals are sent. Charged molecules (ions) build up in the neuron until they reach a point where an electrochemical pulse is triggered, traveling down the axon and across synapses to the dendrites of other neurons. Combine many such events, and we have the makings of a "thought." (We're glossing over some complications here; hopefully neuroscientists will forgive me.)

For the most part, these processes can be thought of as being purely classical, or at least deterministic. Quantum mechanics plays a role at some level in any chemical reaction, since it's quantum mechanics that sets the rules for how electrons want to jump from one atom to another or bind two atoms together. But when you get enough atoms together in one place, their net behavior can be described without any reference to quantum concepts like entanglement or the Born rule—otherwise you wouldn't have been able to take a chemistry class in high school without first learning the Schrödinger equation and worrying about the measurement problem.

So "decisions" are best thought of as classical events, not quantum ones. While you might be personally unsure what choice you will eventually make, the outcome is encoded in your brain. We're not absolutely sure about the extent to which this is true, since there's still a lot we don't know about the physical processes behind thinking. It's possible that the rates of neurologically important chemical reactions can vary slightly depending on the entanglement between the different atoms involved. If that turns out to be true, there would be a sense in which your brain is a quantum computer, albeit a limited one.

At the same time, an honest Everettian admits that there will always be branches of the wave function on which quantum systems appear to have done very unlikely things. As Alice mentioned in Chapter Eight, there will be branches where I run into a wall and happen to tunnel through it, rather than bouncing off. Likewise, even if the classical approximation to my brain implies that I'm going to bet all my chips at the poker table, there is some tiny amplitude for a bunch of neurons to do unlikely things and cause me to make a snug fold. But it's not my decision that's causing the branching; it's the branching that I interpret as leading to my decision.

Under the most straightforward understanding of the chemistry going on in our brains, most of our thinking has nothing to do with entanglement and branching of the wave function. We shouldn't imagine that making a difficult decision splits the world into multiple copies, each containing a version of you that chose differently. Unless, of course, you don't want to take responsibility, and turn your decision-making over to a quantum random-number generator.

○ ○ ○

Similarly, quantum mechanics has nothing to do with the question of free will. It's natural to think that it might, as free will is often contrasted with determinism, the idea that the future is completely determined by the present state of the universe. After all, if the future is determined, what room is there for me to make choices? In the textbook presentation of quantum mechanics, measurement outcomes are truly random, so physics is not deterministic. Maybe that opens the door a crack for free will to sneak back in, after it was banished by the Newtonian clockwork paradigm of classical mechanics?

There's so much wrong with this that it's hard to know where to start. First, "free will" versus "determinism" isn't the right distinction to draw. Determinism should be opposed to "indeterminism," and free will should be opposed to "no free will." Determinism is straightforward to define: given the exact current state of the system, the laws of physics determine precisely the state at later times. Free will is trickier. One usually hears free will defined as something like "the ability to have chosen otherwise." That means we're comparing what really happened (we were in a situation, we made a decision, and we acted accordingly) to a different hypothetical scenario (we wind the clock backward to the original situation, and ask whether we "could have" decided differently). When playing this game, it's crucial to specify exactly what is kept fixed between the real and hypothetical situations. Is it absolutely everything, down to the last microscopic detail? Or do we just imagine fixing our available macroscopic information, allowing for variation within invisible microscopic details?

Let's say we're hard-core about this question, and compare what actually happened to a hypothetical re-running of the universe starting from exactly the same initial condition, down to the precise state of every last elementary particle. In a classical deterministic universe the outcome would be precisely the same, so there's no possibility you could have "made a different decision." By contrast, according to textbook quantum mechanics, an element of randomness is introduced, so we can't confidently predict exactly the same future outcome from the same initial conditions.

But that has nothing to do with free will. A different outcome doesn't mean we manifested some kind of personal, supra-physical volitional influence over the laws of nature. It just means that some unpredictable quantum random numbers came up differently. What matters for the traditional "strong" notion of free will is not whether we are subject to deterministic laws of nature, but whether we are subject to impersonal laws of any sort. The fact that we can't predict the future isn't the same as the idea that we are free to bring it about. Even in textbook quantum mechanics, human beings are still collections of particles and fields obeying the laws of physics.

For that matter, quantum mechanics is not necessarily indeterministic. Many-Worlds is a counterexample. You evolve, perfectly deterministically, from a single person now into multiple persons at a future time. No choices come into the matter anywhere.

On the other hand, we can also contemplate a weaker notion of free will, one that refers to the macroscopically available knowledge we actually have about the world, rather than running thought experiments based on microscopically perfect knowledge. In that case, a different form of unpredictability arises. Given a person and what we (or they, or anyone) know about their current mental state, there will typically be many different specific arrangements of atoms and molecules in their bodies and brains that are compatible with that knowledge. Some of those arrangements may lead to sufficiently different neural processes that we would end up acting very differently, if those arrangements had been true. In that case, the best we can realistically do to describe the way human beings (or other conscious agents) act in the real world is to attribute volition to them—the ability to choose differently.

Attributing volition to people is what every one of us actually does as we go through life talking about ourselves and others. For practical purposes it doesn't matter whether we could predict the future from perfect knowledge of the present, because we don't have such knowledge, nor will we ever. This has led philosophers, going back as far as Thomas Hobbes, to propose *compatibilism* between underlying deterministic laws and the reality of human choice-making. Most modern philosophers are compatibilists about free will (which doesn't mean it's right, of course). Free will is real, just like tables and temperature and branches of the wave function.

As far as quantum mechanics is concerned, it doesn't matter whether you are a compatibilist or an incompatibilist concerning free will. In neither case should quantum uncertainty affect your stance; even if you can't predict the outcome of a quantum measurement, that outcome stems from the laws of physics, not any personal choices made by you. We don't create the world by our actions, our actions are part of the world.

∘ ∘ ∘

I would be remiss to talk about the human side of Many-Worlds without confronting the question of consciousness. There is a long history of claiming that human consciousness is necessary to understand quantum mechanics, or that quantum mechanics may be necessary to understand consciousness. Much of this can be attributed to the impression that quantum mechanics is mysterious, and consciousness is mysterious, so maybe they have something to do with each other.

That's not wrong, as far as it goes. Maybe quantum mechanics and consciousness are somehow interconnected; it's a hypothesis we're welcome to contemplate. But according to everything we currently know, there is no good evidence this is actually the case.

Let's first examine whether quantum mechanics might help us understand consciousness. It's conceivable—though far from certain—that the rates of various neural processes in your brain depend on quantum entanglement in an interesting way, so that they cannot be understood by classical reasoning alone. But accounting for consciousness, as we traditionally think about it, isn't a straightforward matter of the rates of neural processes. Philosophers distinguish between the "easy problem" of consciousness—figuring out how we sense things, react to them, think about them—and the "hard problem"—our subjective, first-person experience of the world; what it is like to be us, rather than someone else.

Quantum mechanics doesn't seem to have anything to do with the hard problem. People have tried: Roger Penrose, for example, has teamed with anesthesiologist Stuart Hameroff to develop a theory in which objective collapse of the wave functions of microtubules in the brain helps explain why we experience consciousness. This proposal has not gained much acceptance in the neuroscience community. More important, it's unclear why it should matter for consciousness at all. It's perfectly conceivable that some subtle quantum processes in the brain, involving microtubules or something completely different, affect the rate at which our neurons fire. But this is of no help whatsoever in bridging the gap between "the firing of our neurons" and "our subjective, self-aware experience." Many scientists and philosophers, myself included, have no trouble believing that this gap is very bridgeable. But a tiny change in the rate of this or that neurochemical process doesn't seem to be relevant to understanding how. (And if it were, there's no reason the effect couldn't be repeated in nonhuman computers.)

Everettian quantum mechanics has nothing specific to say about the hard problem of consciousness that wouldn't be shared by any other view in which the world is entirely physical. In such a view, the relevant facts about consciousness include these:

1. Consciousness arises from brains.
2. Brains are coherent physical systems.

That's all. ("Coherent" here means "made of mutually interacting parts"; two collections of neurons on two non-interacting branches of the wave function are two distinct brains.) You can extend "brains" to "nervous systems" or "organisms" or "information-processing systems" if you like. The point is that we aren't making extra assumptions about consciousness or personal identity in order to discuss Many-Worlds quantum mechanics; it is a quintessentially mechanistic theory, with no special role for observers or experiences. Conscious observers branch along with the rest of the wave function, of course, but so do rocks and rivers and clouds. The challenge of understanding consciousness is as difficult, no more and no less, in Many-Worlds as it would have been without quantum mechanics at all.

There are many important aspects of consciousness that scientists don't currently understand. That is precisely what we should expect; the human mind generally, and consciousness in particular, are extremely complex phenomena. The fact that we don't fully understand them shouldn't tempt us into proposing entirely new laws of fundamental physics to help ourselves out. The laws of physics are enormously better understood, and that understanding has been much better verified by experiment, than the functioning of our brains and their relationship to our minds. We might someday have to contemplate modifying the laws of physics to successfully account for consciousness, but that should be a move of last resort.

° ° °

We can also flip the question on its head: If quantum mechanics doesn't help account for consciousness, is it nevertheless possible that consciousness plays a central role in accounting for quantum mechanics?

Many things are possible. But there's a bit more to it than that. Given the prominence afforded to the act of measurement in the rules of standard textbook quantum theory, it's natural to wonder whether there isn't something special about the interaction between a conscious mind and a quantum system. Could the collapse of the wave function be caused by the conscious perception of certain aspects of physical objects?

According to the textbook view, wave functions collapse when they are measured, but what precisely constitutes "measurement" is left a little vague. The Copenhagen interpretation posits a distinction between quantum and classical realms, and treats measurement as an interaction between a classical observer and a quantum system. Where we should draw the line is hard to specify. If we have a Geiger counter observing emission from a radioactive source, for example, it would be natural to treat the counter as part of the classical world. But we don't have to; even in Copenhagen, we could imagine treating Geiger counters as quantum systems that obey the Schrödinger equation. It's only when the outcome of a measurement is perceived by a human being that (in this way of thinking) the wave function absolutely has to collapse, because no human being has ever reported being in a superposition of different measurement outcomes. So the last possible place we can draw the cut is between "observers who can testify as to whether they are in a superposition" and "everything else." Since the perception of not being in a superposition is part of our consciousness, it's not crazy to ask whether it's actually consciousness that causes the collapse.

This idea was put forward as early as 1939, by Fritz London and Edmond Bauer, and later gained favor with Eugene Wigner, who won the Nobel Prize for his work on symmetries. In Wigner's words:

> All that quantum mechanics purports to provide are probability connections between subsequent impressions (also called "apperceptions") of the consciousness, and even though the dividing line between the observer, whose consciousness is being affected, and the observed physical object can be shifted towards the one or the other to a considerable degree, it cannot be eliminated. It may be premature to believe that the present philosophy of quantum mechanics will remain a permanent feature of future physical theories; it will remain remarkable, in whatever way our future concepts may develop, that the very study of the external world led to the conclusion that the content of the consciousness is an ultimate reality.

Wigner himself later changed his mind about the role of consciousness in quantum theory, but others have taken up the torch. It's not generally a view you will hear spoken of approvingly at physics conferences, but there are some scientists out there who continue to take it seriously.

If consciousness did play a role in the quantum measurement process, what exactly would that mean? The most straightforward approach would be to posit a *dualist* theory of consciousness, according to which "mind" and "matter" are two distinct, interacting categories. The general idea would be that our physical bodies are made of particles with a wave function that obeys the Schrödinger equation, but that consciousness resides in a separate immaterial mind, whose influence causes wave functions to collapse upon being perceived. Dualism has waned in popularity since its heyday in the time of René Descartes. The basic conundrum is the "interaction problem": How do mind and matter interact with each other? In the present context, how is an immaterial mind, lacking extent in space and time, supposed to cause wave functions to collapse?

There is another strategy, however, that seems at once less clunky and considerably more dramatic. This is *idealism*, in the philosophical sense of the word. It doesn't mean "pursuing lofty ideals," but rather that the fundamental essence of reality is mental, rather than physical, in character. Idealism can be contrasted with physicalism or materialism, which suggest that reality is fundamentally made of physical stuff, and minds and consciousness arise out of that as collective phenomena. If physicalism claims that there is only the physical world, and dualism claims that there are both physical and mental realms, idealism claims that there is only the mental realm. (There is not a lot of support on the ground for the remaining logical possibility, that neither the physical nor the mental exists.)

For an idealist, mind comes first, and what we think of as "matter" is a reflection of our thoughts about the world. In some versions of the story, reality emerges from the collective effort of all the individual minds, whereas in others, a single concept of "the mental" underlies both individual minds and the reality they bring to be. Some of history's greatest philosophical minds, including many in various Eastern traditions but also Westerners such as Immanuel Kant, have been sympathetic to some version of idealism.

It's not hard to see how quantum mechanics and idealism might seem like a good fit. Idealism says that mind is the ultimate foundation of reality, and quantum mechanics (in its textbook formulation) says that properties like position and momentum don't exist until they are observed, presumably by someone with a mind.

All varieties of idealism are challenged by the fact that, aside from the contentious exception of quantum measurement, the real world seems to move along quite well without any particular help from conscious minds. Our minds discover things about the world through the process of observation and experiment, and different minds end up discovering aspects of the world that always end up being wholly consistent with one another. We have assembled quite a detailed and successful account of the first few minutes of the history of the universe, a time when there were no known minds around to think about it. Meanwhile, progress in neuroscience has increasingly been able to identify particular thought processes with specific biochemical events taking place in the material that makes up our brains. If it weren't for quantum mechanics and the measurement problem, all of our experience of reality would speak to the wisdom of putting matter first and mind emergent from it, rather than the other way around.

So, is the weirdness of the quantum measurement process sufficiently intractable that we should discard physicalism itself, in favor of an idealistic philosophy that takes mind as the primary ground of reality? Does quantum mechanics necessarily imply the centrality of the mental?

No. We don't *need* to invoke any special role for consciousness in order to address the quantum measurement problem. We've seen several counterexamples. Many-Worlds is an explicit example, accounting for the apparent collapse of the wave function using the purely mechanistic process of decoherence and branching. We're allowed to contemplate the possibility that consciousness is somehow involved, but it's just as certainly not forced on us by anything we currently understand. Of course, we will often talk about conscious experiences in our attempts to map the quantum formalism onto the world as we see it, but only when the things we're trying to explain are those experiences themselves. Otherwise, minds have nothing to do with it.

These are difficult, subtle issues, and this isn't the place for a completely fair and comprehensive adjudication of the debate between idealism and physicalism. Idealism isn't something that's easy to disprove; if someone is convinced it's right, it's hard to point to anything that would obviously change their mind (or Mind). But what they can't do is claim that quantum mechanics forces us into such a position. We have very straightforward and compelling models of the world in which reality exists independently of us; there's no need to think we bring reality into existence by observing or thinking about it.

Part Three

SPACETIME

11

Why Is There Space?

Emergence and Locality

Okay, at long last we're ready to think about the actual world.

Wait a minute, I hear you thinking. *I thought we were talking about the actual world already. Isn't quantum mechanics supposed to describe the actual world?*

Well, sure. But quantum mechanics can also describe plenty of worlds other than our actual one. Quantum mechanics itself isn't a single theory, in the sense of being a model of one specific physical system. It's a framework, just like classical mechanics is, in which we can talk about many different physical systems. We can talk about the quantum theory of a single particle, or of the electromagnetic field, or of a set of spins, or of the entire universe. Now it's time to focus in on what the quantum theory of our actual world might look like.

This goal—finding the right quantum theory of the actual world—has been pursued by generations of physicists since the early twentieth century. By any possible measure, they have been extraordinarily successful. One important insight was to think of the basic building blocks of nature not as particles but as fields pervading space, thus leading to *quantum field theory.*

Back in the nineteenth century, physicists seemed to be homing in on a view of the world in which both particles and fields played a role: matter was made of particles, and the forces by which they interacted were described by fields. These days we know better; even the particles that we know and love are actually vibrations in fields that suffuse the space around us. When we see particle-like tracks in a physics experiment, that's a reflection of the fact that what we see is not what there really is. Under the right circumstances we see particles, but our best current theories say that fields are more fundamental.

Gravity is the one part of physics that doesn't fit comfortably into the quantum-field-theory paradigm. You will often hear that "we don't have a quantum theory of gravity," but that's a bit too strong. We have an extremely good classical theory of gravity: Einstein's general relativity, which describes the curvature of spacetime. General relativity is itself a field theory—it describes a field pervading all of space, in this case the gravitational field. And we have very well understood procedures for taking a classical field theory and quantizing it, yielding a quantum field theory. Apply those procedures to the known fields of fundamental physics, and we end up with something called the *Core Theory.* The Core Theory accurately describes not only particle physics but also gravity, as long as the strength of the gravitational field doesn't grow too large. It is sufficient to describe every phenomenon that happens in your everyday experience, and quite a bit beyond—tables and chairs, amoebas and kittens, planets and stars.

The problem is that the Core Theory doesn't cover a number of situations beyond the everyday, including places where gravity becomes extreme, like black holes and the Big Bang. In other words, we have a theory of quantum gravity that is adequate when gravity is fairly weak, one that is perfectly capable of describing why apples fall from trees or how the moon orbits the Earth. But it's limited; once gravity becomes very strong, or we try to push our calculations too far, our theoretical apparatus fails us. As far as we can tell, this situation is unique to gravity. For all the other particles and forces, quantum field theories seem to be able to handle any situation we can imagine.

Faced with the difficulty of quantizing general relativity as we would any other field theory, there are a number of strategies that we might try. One is simply to think harder; maybe there is a good way to directly quantize general relativity, but it involves new techniques that we haven't needed for other field theories. A different approach is to imagine that general relativity isn't the right theory to quantize; maybe we should start with a distinct classical precursor, such as string theory, and then quantize that, hoping to build a quantum theory that includes gravity along with everything else. Physicists have been trying both of these approaches for some decades now, with some successes but still a lot of puzzles left unanswered.

Here we're going to consider a different strategy, one that faces up to the quantum nature of reality from the start. Every physicist understands that the world is fundamentally quantum, but as we actually do physics we can't help but be influenced by our experience and intuitions, which have long been trained on classical principles. There are particles, there are fields, they do things, we can observe them. Even when we explicitly move to quantum mechanics, physicists generally start by taking a classical theory and quantizing it. But nature doesn't do that. Nature simply *is* quantum from the start; classical physics, as Everett insisted, is an approximation that is useful in the right circumstances.

This is where we reach the payoff for all of our hard work over the previous chapters. Many-Worlds is uniquely suited to the task of throwing away all of our classical intuition, being quantum from the get-go, and determining how the approximately classical world that we see around us ultimately emerges from the wave function of the universe, spacetime and all.

In alternatives to Many-Worlds, one often needs additional variables (such as in Bohmian mechanics) or rules about how wave functions spontaneously collapse (such as in GRW). These are typically derived from our experience with the classical limit of the theory under consideration, and it's exactly that experience that has failed us so far for quantum gravity. Many-Worlds, by contrast, doesn't rely on any additional superstructure. Ultimately it's not a theory of particular kinds of "stuff," just quantum states evolving under the Schrödinger equation. That creates extra work for us under ordinary circumstances, as we have to explain why we see a world of particles and fields at all. But in this unique quantum-gravity context, it's an advantage, since we have to do that work anyway. Many-Worlds, with its quantum-first perspective, is the right approach if you feel that we don't know of any classical theory that could serve as the right starting point for constructing a quantum theory of gravity.

∘ ∘ ∘

Before digging into quantum gravity proper, we need to lay some groundwork. General relativity is a theory of the dynamics of spacetime, so in this chapter we'll ask why the concept of "space" is so important in the first place. The answer resides in the concept of *locality*—things interact with one another when they are nearby in space. In the next chapter we'll see how quantum fields propagating through space embody this principle of locality, and teach us something about the nature of empty space. In the chapter after that we'll investigate how to extract space itself from the quantum wave function. And in the final chapter we'll see that when gravity becomes strong, locality itself will have to be abandoned as a central principle. The mystery of quantum gravity seems to be intimately connected with the virtues and the shortcomings of the idea of locality.

It's worth being careful about "locality," as it is used in two somewhat different senses: what we might call *measurement locality* and *dynamical locality*. The EPR thought experiment shows that there is something that seems nonlocal about quantum measurement. Alice measures her spin, and what Bob will measure for his spin far away is immediately affected, even if he doesn't know it. Bell's theorem implies that any theory in which measurements have definite outcomes—basically, every approach to quantum mechanics other than Many-Worlds—is going to feature this kind of measurement nonlocality. Whether Many-Worlds is nonlocal in this sense depends on how we choose to define our branches of the wave function; we're allowed to make either local or nonlocal choices, where branching happens only nearby or immediately all throughout space.

Dynamical locality, on the other hand, refers to the smooth evolution of the quantum state when no measurement or branching is happening. That's the context in which physicists expect everything to be perfectly local, with disturbances at one location only immediately affecting things right nearby. This kind of locality is enforced by the rule in special relativity that nothing can travel faster than light. And it's this dynamical locality that we're concerned with at the moment as we study the nature and emergence of space itself.

With that in mind, we can roll up our sleeves a bit and dig into the question of how the structure of our observed reality—we live in a world that looks like a collection of objects located in space, behaving approximately classically except for occasional quantum jumps—emerges from the quantum wave function. Everettian quantum mechanics purports to tell a story about many such worlds, but the postulates of the theory (wave functions, smooth evolution) don't even mention "worlds" at all. Where do the worlds come from, and why do worlds look approximately classical?

In our discussion of decoherence, we pointed out that you can think of a quantum system as having split into multiple separate copies once it becomes entangled with the larger environment around it, since whatever happens to each copy won't be able to interfere with whatever happens to the others. If we want to be sticklers, however, that's telling us that we're *allowed* to think about the decohered wave function as describing separate worlds—not that we *should* think of it that way, much less that we *need* to think of it that way. Can we do better?

The truth is, nothing forces us to think of the wave function as describing multiple worlds, even after decoherence has occurred. We could just talk about the entire wave function as a whole. It's just really helpful to split it up into worlds.

Many-Worlds describes the universe using a single mathematical object, the wave function. There are many ways of *talking about* the wave function that give us physical insight into what is going on. It may be useful in some cases to talk in terms of position, for example, and in other cases in terms of momentum. Likewise, it is often helpful to talk about the post-decoherence wave function as describing a set of distinct worlds; that's justified, because what happens on each branch doesn't affect what happens on the others. But ultimately, that language is a convenience for us, not something that the theory itself insists on. Fundamentally, the theory just cares about the wave function as a whole.

By way of an analogy, think of all the matter in the room around you right now. You could describe it—helping ourselves to the classical approximation for the moment—by listing the position and velocity of every atom in the room. But that would be crazy. You neither have access to all that information, nor could you put it to use if you did, nor do you really need it. Instead, you chunk up the stuff around you into a set of useful concepts: chairs, tables, lights, floors, and so on. That's an enormously more compact description than listing every atom would be, but still gives us a great deal of insight into what's going on.

Similarly, characterizing the quantum state in terms of multiple worlds isn't necessary—it just gives us an enormously useful handle on an incredibly complex situation. As Alice insisted in Chapter Eight, the worlds aren't fundamental. Rather, they're *emergent*.

Emergence in this sense does not refer to events unfolding over time, as when a baby bird emerges from its egg. It's a way of describing the world that isn't completely comprehensive, but divides up reality into more manageable chunks. Notions like rooms and floors are nowhere to be found in the fundamental laws of physics—they're emergent. They are ways of effectively describing what's going on even if we lack perfect knowledge of each and every atom and molecule around us. To say that something is emergent is to say that it's part of an approximate description of reality that is valid at a certain (usually macroscopic) level, and is to be contrasted with "fundamental" things, which are part of an exact description at the microscopic level.

In the Laplace's demon thought experiment, we imagine a vast intelligence that would know all the laws of physics and the exact state of the world, as well as having unlimited computational capacity. To the demon, everything that is, was, and ever will be is completely known. But none of us is Laplace's demon. In reality, we have at best partial information about the state of the world, and quite limited computational capacity. None of us looks at a cup of coffee and sees every particle in every atom; we see some coarse macroscopic features of the liquid and the cup. But that can be all the information we need to have a useful discussion about the coffee, and to predict its behavior in a variety of circumstances. A cup of coffee is an emergent phenomenon.

The same thing can be said for worlds in Everettian quantum mechanics. For a quantum version of Laplace's demon, with exact knowledge of the quantum state of the universe, there would never be any need to divide the wave function into a set of branches describing a collection of worlds. But it is enormously convenient and helpful to do so, and we're allowed to take advantage of this convenience because the individual worlds don't interact with one another.

That doesn't mean that the worlds aren't "real." Fundamental versus emergent is one distinction, and real versus not-real is a completely separate one. Chairs and tables and cups of coffee are indubitably real, as they describe true patterns in the universe, ones that organize the world in ways that reflect the underlying reality. The same goes for Everettian worlds. We choose to invoke them when carving up the wave function for our convenience, but we don't do that carving randomly. There are right and wrong ways to divide the wave function into branches, and the right ways leave us with independent worlds that obey approximately classical laws of physics. Which ways actually work is ultimately determined by the fundamental laws of nature, not by human whimsy.

∘ ∘ ∘

Emergence is not a generic feature of physical systems. It happens when there's a special way of describing the system that involves much less information than a complete description would, but nevertheless gives us a useful handle on what's going on. That's why it makes sense for us to carve up reality in the way we do, describing tables and chairs and branches of the wave function.

Think of a planet orbiting the sun. A planet like the Earth contains roughly 10^{50} particles. To describe the state of the Earth exactly, even at the classical level, would require listing the position and momentum of every one of those particles, something that is beyond even our wildest imagination of supercomputing power. Happily, if what we care about is just the orbit of the planet, the vast majority of that information is completely unnecessary. We can instead idealize the Earth as a single point, located at the Earth's center of mass and with the same total momentum. The state of this idealized point is specified by a position and momentum, and that very tiny amount of information (six numbers, three each for position and momentum, as opposed to 6×10^{50} numbers, positions and momenta for each particle) is all we need to calculate its trajectory. That's emergence: a way of capturing important features of a system using far less information than an exhaustive description would entail.*

We often talk about emergent descriptions in terms of how "convenient" they are for us to use, but don't be tricked into thinking there's anything anthropocentric going on. Tables and chairs and planets would still exist even if there were no human beings to talk about them. "Convenience" is a shorthand for indicating an objective physical property: the existence of an accurate model of the system that requires only a tiny fraction of the full information characterizing it.

Emergence is not automatic. It's a special, precious thing, and provides an enormous simplification when it occurs. Imagine we know the position of every one of the 10^{50} particles in the Earth, but we don't know the momentum of any of them. We possess an enormous amount of information—fully half of the total information available—but we have precisely zero ability to predict where the Earth would be going next. Strictly speaking, even if we know the momentum of all but one of the particles in the Earth, but have no knowledge at all of exactly one momentum, we can't say what the Earth will do next; it's possible that this single particle has as much momentum as all of the others combined.

That's the generic situation in physics. In order to accurately predict what a system made of many parts will do next, you need to keep track of the information of all the parts. Lose just a little bit, and you know nothing. Emergence happens when the opposite is possible: we can throw away almost all the information, keeping just a little bit (as long

* Sadly there are competing definitions of the word "emergence," some of which mean almost the opposite of the sense used here. Our definition is sometimes called "weak emergence" in the literature, as opposed to "strong emergence," in which the whole is irreducible to the sum of its parts.

as you correctly identify which bit), and still say quite a lot about what will happen.

In the case of the center of mass of an object made of many particles, the kind of information in the emergent description we have is exactly the same as the kind we started with (position and momentum), just a lot less of it. But emergence can be more subtle than that; the emergent description may be of an entirely different thing from what we started with.

Consider the air in our room. Imagine that we divide space into tiny boxes, perhaps one millimeter on each side. Each box still contains a huge number of molecules. But instead of keeping track of the state of each one of them, we keep track of average quantities such as the density, pressure, and temperature in each box. It turns out that this is all the information we need to make accurate predictions for how the air will behave. The emergent theory describes a different kind of thing, a fluid rather than a collection of molecules, but that fluid description suffices to describe the air to a high degree of precision. Treating the air as a fluid requires much less data than treating it as a collection of particles; the fluid description is emergent.

Everettian worlds are the same way. We don't need to keep track of the entire wave function to make useful predictions, just what happens in an individual world. To a good approximation we can treat what happens in each world using classical mechanics, with just the occasional quantum intervention when we entangle with microscopic systems in superposition. That's why Newton's laws of gravitation and motion are sufficient to fly rockets to the moon without knowing the complete quantum state of the universe; our individual branch of the wave function describes an emergent almost-classical world.

Branches of the wave function, describing separate worlds, are not mentioned in the postulates of Many-Worlds. Nor are tables and chairs and air mentioned in the Core Theory of particles and forces. As the philosopher Daniel Dennett has put it, in terms that were then ported into the quantum context by David Wallace, each world is an emergent feature that captures "real patterns" within the underlying dynamics. A real pattern gives us an accurate way of talking about the world, without appealing to a comprehensive microscopic description. That's what makes emergent patterns in general, and Everettian worlds in particular, indisputably real.

o o o

Once you believe that branches of the wave function can usefully be thought of as emergent worlds, you might start wondering why it's this set of worlds in particular. Why do we end up seeing macroscopic objects with pretty well-defined locations in space, rather than being in superpositions of different locations? Why is "space" apparently such a central concept at all? Textbooks in introductory quantum mechanics sometimes give the impression that classical behavior is inevitable once objects become very big, but that's nonsense. We have no trouble at all imagining a wave function that describes macroscopic objects in all sorts of weird superpositions. The real answer is more interesting.

We can begin to get a handle on the special nature of space by comparing how we think about position to how we think about momentum. When Isaac Newton first wrote down the equations of classical mechanics, position clearly played a privileged role, whereas velocity and momentum were derived quantities. Position is "where you are in space," while velocity is "how fast you are moving through space," and momentum is mass times velocity. Space would appear to be the main thing.

But a deeper look reveals that the concepts of position and momentum are on more of an equal footing than they first appear. Perhaps we shouldn't be surprised; after all, position and momentum are the two quantities that together define the state of a classical system. Indeed, in the Hamiltonian formulation of classical mechanics, position and momentum are explicitly on an equal footing. Is this a reflection of some underlying symmetry that isn't obvious on the surface?

In our everyday lives, position and momentum seem quite different. What a mathematician would call "the space of all possible positions" is what the rest of us just call "space"; it's the three-dimensional world in which we live. The "space of all possible momenta," or "momentum space," is also three-dimensional, but it's a seemingly abstract concept. Nobody believes we live there. Why not?

The feature that makes space special is locality. Interactions between different objects happen when they are nearby in space. Two billiard balls bounce off each other when they come together at the same spatial position. Nothing of the sort happens when particles have the same (or opposite) momenta; if they're not in the same location, they just keep going their merry way. That's not a necessary feature of the laws of physics—we could imagine other possible worlds where it wasn't the case—but it's one that seems to hold pretty well in our world.

Ricocheting billiard balls are classical, but the same discussion could be had about quantum mechanics. The basic quantum formalism also treats position and momentum equally. We can express the wave function by attaching a complex amplitude to every possible location the particle can be in, or we could just as well express it by attaching a complex number to every possible momentum the particle could have. The two ways of describing the same underlying quantum state are equivalent, expressing the same information in different ways, as we saw when discussing the uncertainty principle.

This is kind of profound. We've said that a wave function of definite momentum looks like a sine wave. But that's what it looks like in terms of position, which is the language we naturally tend to speak. Expressed in terms of momentum, the same quantum state would look like a spike located at that particular momentum. A state with definite position would look like a sine wave spread over all possible momenta. This begins to suggest that what really matters is the abstract notion of "the quantum state," not its specific realization as a wave function in terms of either position or momentum.

The symmetry is broken, once again, by the fact that in our particular world, interactions happen when systems are nearby in space. This is dynamical locality at work. From a Many-Worlds perspective that treats quantum states as fundamental and everything else as emergent, this suggests that we should really turn things around: "positions in space" are the variables in which interactions look local. Space isn't fundamental; it's just a way to organize what's going on in the underlying quantum wave function.

○ ○ ○

This point of view helps us understand why the Everettian wave function can naturally be divided into a set of approximately classical worlds. This issue is known as the *preferred-basis problem*. Many-Worlds is based on the fact that the wave function of the universe will generally describe all sorts of superpositions, including states where macroscopic objects are in superpositions of being in very different locations. But we never see chairs or bowling balls or planets in superpositions; as far as our experience is concerned, they always seem to have definite locations, and their motion obeys the rules of classical mechanics to a very good approximation. Why don't the states we see ever involve macroscopic superpositions? We can write the wave function as a combination of many distinct worlds, but why divide it up into *these* worlds in particular?

The answer was essentially figured out in the 1980s, using decoherence, although researchers are still hammering out the details. To get there, it's useful to turn to that old thought-experiment standby, Schrödinger's Cat. We have a sealed box containing a cat and a container of sleeping gas. Schrödinger's original scenario involved poison, but there's no reason we have to imagine killing the cat. (His daughter Ruth once mused, "I think my father just didn't like cats.")

Our experimenter has rigged a spring to pull open the container, releasing the gas and putting the cat to sleep, but only when a detector such as a Geiger counter clicks upon detecting a particle of radiation. Next to the detector is a radioactive source. We know the rate at which particles are emitted from the source, so we can calculate the probability that the counter will click and release the hammer after any given period of time.

Radioactive emission is a fundamentally quantum process. What we informally describe as the occasional, random emission of a particle is actually a smooth evolution of the wave function of the atomic nuclei within the source. Each nucleus evolves from a state of purely undecayed to a superposition of (un-decayed)+(decayed), with the latter part gradually growing over time. The emission appears random because the detector doesn't measure the wave function directly; it only sees either (un-decayed) or (decayed), just as a vertical Stern-Gerlach magnet only ever sees spin-up or spin-down.

The point of the thought experiment is to take a microscopic quantum superposition and magnify it to a manifestly macroscopic situation. That happens as soon as the detector clicks. All the business with the sleeping gas and the cat is just to make the amplification of a quantum superposition to the macroscopic world more vivid. (The word "entanglement," or in German *Verschränkung*, was first applied to quantum mechanics by Schrödinger in the discussion of his cat, which arose out of correspondence with Einstein.)

Schrödinger's experiment was posed in the context of the textbook approach to the measurement problem, where wave functions collapse when they are literally observed. So, he says, imagine that we keep the box closed—not observing what's inside—until the wave function evolves to an even superposition of "at least one nucleus has decayed" and "no nuclei have decayed." In that case, the wave functions of the detector, the gas, and the cat will all also evolve into an equal superposition, of "the detector clicked, the gas was released, and the cat is asleep" and "the detector didn't yet click, the gas is still in the container, and the cat is awake." Surely, asks Schrödinger, you don't seriously believe that the box contains a superposition of an awake cat and an asleep cat until we open it?

As far as that goes, he was right. Once we have an Everettian perspective on quantum dynamics, we accept that the wave function smoothly evolves into an equal superposition of two possibilities, one in which the cat is asleep and the other in which it is awake. But decoherence tells us that the cat is also entangled with its environment, consisting of all the air molecules and photons within the box. The effective branching into separate worlds happens almost right away after the detector clicks. By the time the experimenter gets around to opening the box, there are two branches of the wave function, each of which has a single cat and a single experimenter, not a superposition.

This solves Schrödinger's original worry, but raises another one. Why is it that when we open the box, the particular decohered quantum states we see are either that of an awake cat, or an asleep cat? Why don't we see some superposition of both? "Awake" and "Asleep" together represent just one possible basis for the cat system, just as "spin-up" and "spin-down" do for the electron. Why is that basis preferred over any other one?

The physical process that matters is stuff in the environment—gas molecules, photons—interacting with the physical system under consideration. Whether a particular particle actually does interact with the cat will depend on where the cat is. A given photon might very well be absorbed by a cat that is awake and prowling around the box, but completely miss a cat that is sleeping on the floor.

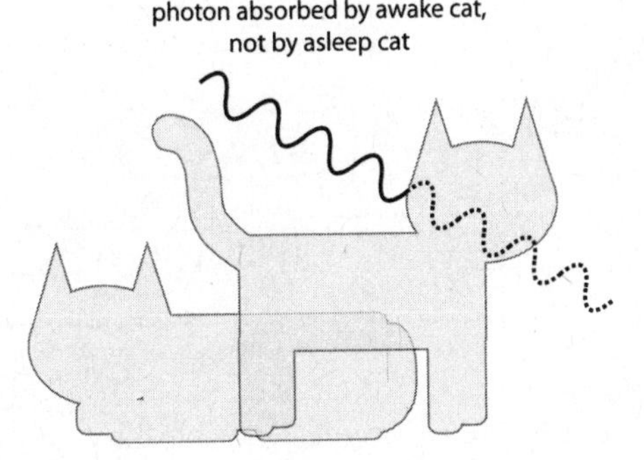

What's special about the "Awake"/"Asleep" basis, in other words, is that the individual states describe well-defined configurations in space. And space is the quantity with respect to which physical interactions are local. A particle can bump into a cat if the particle and the cat come into physical contact. The two parts of the cat wave function, "Awake" and "Asleep," come into contact with different particles in the environment, and therefore branch into different worlds.

This is the basic answer to the question of why we see the particular worlds that we do: the preferred-basis states are those that describe coherent objects in space, because such objects interact consistently with their environments. These are often called *pointer states*, as they are the states in which the pointer of a macroscopic measuring device will indicate a definite value, rather than being in a superposition. The pointer basis is where a well-behaved classical approximation makes sense, and therefore it's that kind of basis that defines emergent worlds. Decoherence is the phenomenon that ultimately links the austere simplicity of Everettian quantum mechanics to the messy particularity of the world we see.

12 A World of Vibrations

Quantum Field Theory

The phrase "action at a distance," usually modified by Einstein's adjective "spooky," is often invoked in discussions of quantum entanglement and the EPR puzzle. But the idea is much older than that—it goes back at least to Isaac Newton and his theory of gravity.

If Newton had done nothing more than put together the basic structure of classical mechanics, he would be a leading candidate for the greatest physicist of all time. What clinches his claim to the crown is that he did much more than that, including little things like inventing calculus. Still, when most people see a picture of Newton in his magnificent wig, they think of his theory of gravity.

Newtonian gravity can be summed up in the famous inverse-square law: the gravitational force between two objects is proportional to the mass of each of them, and inversely proportional to the square of the distance between them. So if you moved the moon to be twice as far away from the Earth, the gravitational force between them would be only one-fourth as large. Using this simple rule, Newton was able to show that planets would naturally move in ellipses around the sun, confirming the empirical relationship that had been posited by Johannes Kepler years before.

But Newton was never really satisfied with his own theory, precisely because it featured action at a distance. The force between two objects depends on where each of them is located, and when an object moves, the direction of its gravitational pull changes instantaneously all throughout the universe. There was nothing in between that would mediate such a change; it simply happened. This bugged Newton—not because it was illogical or incompatible with observation, but just because it seemed wrong. Spooky, one might say.

> It is inconceivable that inanimate brute matter should, without the Mediation of something else which is not material, operate upon and affect other matter without mutual contact. . . . Gravity must be caused by an agent acting constantly according to certain laws; but whether this agent be material or immaterial, I have left to the consideration of my readers.

There is indeed an "agent" that causes gravity to act the way it does, and that agent is perfectly material—it's the gravitational field. This concept was first introduced by Pierre-Simon Laplace, who was able to rewrite Newton's theory of gravity so that the force was carried by a gravitational potential field, rather than simply hopping mysteriously across infinite distances. But a change in the force still happened instantaneously through all of space. It wasn't until Einstein came along with general relativity that changes in the gravitational field, just like changes in the electromagnetic field, were shown to travel through space at the speed of light. General relativity replaces Laplace's potential with the "metric" field, a mathematically sophisticated way of characterizing the curvature of spacetime, but the general idea of a gravitational field pervading all of space has remained intact.

The idea of a field carrying a force is conceptually appealing because it instantiates the idea of locality. As the Earth moves, the direction of its gravitational pull doesn't change instantly throughout the universe. Rather, it changes right where the Earth is located, and then the field at that point tugs on the field nearby, which tugs on the field a little farther away, and so on in a wave moving outward at the speed of light.

Modern physics extends this idea to literally everything in the universe. The Core Theory is constructed by starting with a set of fields and then quantizing them. Even particles like electrons and quarks are really vibrations in quantum fields. That's a wonderful story all by itself, but our aim in this chapter is slightly more modest: to understand the "vacuum" in quantum field theory, the quantum state corresponding to empty space. (I've relegated a brief discussion of states with actual particles in them to the Appendix.) Later we'll tackle the quantum emergence of space itself, but for now we'll be drearily conventional and think about quantum field theory as what you get when you quantize a classical field theory in a preexisting space.

One of the lessons we will learn is that entanglement plays an even more central role in quantum field theory than it does in quantum particle theories. When particles were our primary concern, entanglement was something that may or may not have been important, depending on the physical circumstances. You can create a state of two entangled electrons, but there are plenty of interesting states of two electrons where the particles aren't entangled at all. In field theory, by contrast, essentially every physically interesting state is one that features an enormous amount of entanglement. Even empty space, which you might think of as pretty straightforward, is described in quantum field theory as an intricate collection of entangled vibrations.

∘ ∘ ∘

Quantum mechanics first began when Planck and Einstein argued that electromagnetic waves had particle-like properties, and then Bohr, de

Broglie, and Schrödinger suggested that particles could have wave-like aspects. But there are two different kinds of "waviness" at work here, and it's worth being careful to distinguish between them. One kind of waviness arises when we make the transition from a classical theory of particles to a quantum version, obtaining the quantum wave function of a set of particles. The other kind is when we have a classical field theory to start with, even before quantum mechanics becomes involved at all. That's the case with classical electromagnetism, or with Einstein's theory of gravity. Classical electromagnetism and general relativity are both theories of fields (and therefore of waves), but are themselves perfectly classical.

In quantum field theory, we start with a classical theory of fields and construct a quantum version of *that*. Instead of a wave function that tells us the probability of seeing a particle at some location, we have a wave function that tells us the probability of seeing a particular configuration of a field throughout space. A wave function of a wave, if you like.

There are many ways to quantize a classical theory, but the most direct one is the route we have already taken. Thinking of a collection of particles, we can ask, "Where can the particles be?" The answer for each individual particle is simply "At any point in space." If there were just one particle, the wave function would therefore assign an amplitude to every point in space. But when we have several particles, there isn't a separate wave function for each particle. There is one big wave function, assigning a different amplitude to every possible set of locations that all the particles could be in at once. That's how entanglement can happen; for every configuration of the particles, there is an amplitude we could square to get the probability of observing them there all at the same time.

It's the same thing for fields, with "possible configuration of the particles" replaced by "possible configurations of the field," where by "configuration" we now mean the values of the field at each point

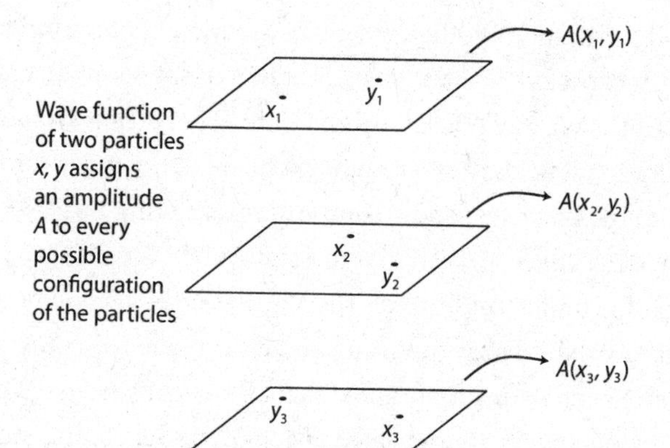

throughout all of space. This wave function considers every possible field configuration, and assigns an amplitude to each. If we could imagine observing the field everywhere at once, the probability of getting any particular shape of the field will be equal to the square of the amplitude assigned to that configuration.

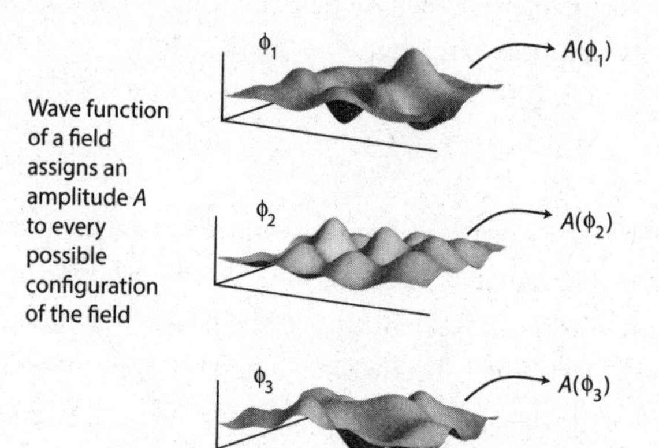

This is the difference between a classical field and a quantum wave function. A classical field is a function of space, and a classical theory with many fields would describe multiple functions of space overlapping with one another. The wave function in quantum field theory is not a function of space, it's a function of the set of all configurations of all the classical fields. (In the Core Theory, that would include the gravitational field, the electromagnetic field, the fields for the various subatomic particles, and so on.) An intimidating beast, but something physicists have learned to understand and even cherish.

All of this implicitly assumes the Many-Worlds version of quantum mechanics. We didn't say anything about decoherence and branching, but we have been taking for granted that all we really need is a quantum wave function and an appropriate version of the Schrödinger equation, and the rest will take care of itself. That's exactly the Everettian situation. (Sometimes when people say "the Schrödinger equation" they are referring specifically to the version Schrödinger originally wrote down, which is only appropriate for non-relativistic point particles, but there's no difficulty in finding a version of the equation for relativistic quantum fields or any other system with a Hamiltonian.) In other theories, one often needs additional variables or rules about how wave functions spontaneously collapse. When we move to field theory, it's not immediately clear what those extra ingredients should be.

○ ○ ○

If quantum field theory describes the world as a wave function of a classical field configuration, that seems to be waviness on top of waviness. If we asked how much wavier things could possibly get, the answer (to paraphrase Nigel Tufnel of Spinal Tap) might be "none more wavy." And yet, when we make observations of quantum fields, for example, in a detector at the Large Hadron Collider in Geneva, what we see are individual tracks representing the paths of point-like objects, not diffuse wavy clouds. Somehow we have circled back to particles, despite being as wavy as can be.

The reason for this goes back to the same reason why we see discrete energy levels for electrons in atoms. An electron moving through space all by itself can have any energy at all, but in the vicinity of the attractive force exerted by an atomic nucleus, it's as if the electron is trapped in a box. The wave function falls to zero far away from the atom; we can think of it as being tied down, just as for a string tied down on both ends and free to move in between. In such circumstances, the tied-down string can only perform a discrete set of vibrations; likewise, the wave function of the electron has a discrete set of energy levels. Anytime the wave function of a system is "tied down" by going to zero for large/faraway/extreme configurations, it will exhibit a set of discrete energy levels.

Returning to field theory, consider a very simple field configuration, a sine wave stretching throughout all of space. We call such a configuration a *mode* of the field; it's a convenient way of thinking, since any field configuration at all can be thought of as a combination of many modes of different wavelengths. That sine wave contains energy, and the energy increases rapidly as we imagine waves of greater and greater height. We want to construct the quantum wave function *of* that field. Because the energy of the field rises with the height of the wave, the wave function needs to decrease rapidly as the height of the wave increases, so as to not give too much probability to very high-energy waves. For all intents and purposes, the wave function is tied down (it goes to zero) at large energies.

As a result, just like a vibrating string or an electron in an atom, there is a discrete set of energy levels for the vibrations of a quantum field. In fact, every mode of the field can be in its lowest-energy state, or its next-highest, or next-highest, and so on. The overall minimum-energy wave function is one in which every single mode has the lowest possible energy. That's a unique state, which we call the *vacuum*. When quantum field theorists talk about the vacuum, they don't mean a machine that lifts dust off your floors, or even a region of interplanetary space devoid of matter. What they mean is "the lowest-energy state of your quantum field theory."

You might think that the quantum vacuum would be empty and boring, but it's actually a wild place. An electron in an atom has a lowest-energy state it can be in, but if we think about it as a wave function of the position of the electron, that function can still have an interesting shape. Likewise, the vacuum state in field theory can still have interesting structure if we ask about individual parts of the field.

The next energy level has a bit more going on, since we make it out of the next-highest energies of each mode. That gives us a bit of freedom; there can be states that are mostly short-wavelength modes, or states that are mostly long-wavelength modes, or any mixture. What they have in common is each mode is in its "first excited state," with just a bit more energy than the minimum.

Putting that together, the wave function for the first excited state of a quantum field theory looks exactly like that of a single particle, expressed as a function of momentum rather than position. There will generally be contributions from different wavelengths, which we interpret as different momenta in the particle wave function. Most important, this kind of state behaves in a particle-like way when we observe it: if we measure a bit of energy in one location (interpreted as "I just saw a particle there"), it becomes overwhelmingly probable that you will observe the same amount of energy nearby if you look a moment later, even if the wave function was originally all spread out. What you end up seeing is a localized vibration propagating in the field, leaving a track in an experimental detector just like a particle is supposed to do. If it looks like a particle and quacks like a particle, it makes sense to call it a particle.

Can we have a quantum-field-theory wave function that combines some modes in their lowest-energy states and some others in their first excited states? Sure—that would be a superposition of a zero-particle state and a one-particle state, giving a state without a definite number of particles.

As you might be prepared to guess, the next-highest energy wave functions of a quantum field theory look like the wave function of two particles. The story goes on for quantum field states representing three particles, or four, or whatever. Just as we observe Schrödinger's cat to be either awake or asleep, and not any superposition thereof, collections of particles are what we observe when we make measurements of gently vibrating quantum fields. In the language of the previous chapter, as long as the fields aren't fluctuating too wildly, the "pointer states" of quantum field theory look like collections of definite numbers of particles. Those are the kinds of states we see when we actually look at the world.

Even better, quantum field theory can describe transitions between states with different numbers of particles, just as an electron can hop up or down in energy in an atom. In ordinary particle-based quantum mechanics, the number of particles is fixed, but quantum field theory has no problem describing particles decaying or annihilating or being created in collisions. Which is good, because things like that happen all the time.

Quantum field theory represents one of the great triumphs of unification in the history of physics, tying together the seemingly opposed ideas of particles and waves. Once we realize that quantizing the electromagnetic field leads to particle-like photons, perhaps it shouldn't be surprising that other particles such as electrons and quarks also arise from quantized fields. Electrons are vibrations in the electron field, various types of quarks are vibrations in various types of quark fields, and so on.

Introductions to quantum mechanics sometimes contrast particles and waves as if they are two equal sides of the same coin, but ultimately the battle between particles and fields is not a fair fight. Fields are more fundamental; it's fields that provide the best picture we currently have of what the universe is made of. Particles are simply what we see when we observe fields under the right circumstances. Sometimes the circumstances aren't right; inside a proton or neutron, even though we often speak about quarks and gluons as if they're individual particles, it's more accurate to think of them as diffuse fields. As physicist Paul Davies once titled a paper, with only a bit of rhetorical exaggeration, "Particles Do Not Exist."

○ ○ ○

Our interest here is in the basic paradigm of quantum reality, not in the specific pattern of particles and their masses and interactions. We care about entanglement and emergence and how the classical world arises from the branching wave function. Happily, for these purposes we can concentrate our attention on the quantum field theory vacuum—the physics of empty space, without any particles flying around.

To bring home the interestingness of the field-theory vacuum, let's focus on one of its most obvious aspects, its energy. It's tempting to think that the energy is zero by definition. But we've been careful not to say that: the vacuum is the "lowest-energy state," not necessarily a "zero-energy state." In fact, its energy can be anything at all; it's a constant of nature, a parameter of the universe that is not determined by any other set of measurable parameters. As far as quantum field theory is concerned, you have to just go out and measure what the energy of the vacuum actually is.

And we have measured the vacuum energy, or at least we think we have. It's not easy to do; you can't simply put a cupful of empty space on a scale and ask how much it weighs. The way to do it is to look for the gravitational influence of the vacuum energy. According to general relativity, energy is the source of the curvature of spacetime, and therefore of gravity. The energy of empty space takes a particular form: there is a precisely constant amount in every cubic centimeter of space, unchanging through the universe, even as spacetime expands or warps. Einstein referred to the vacuum energy as the *cosmological constant*, and cosmologists long debated whether its value was exactly zero or some other number.

That debate seems to have been settled in 1998, when astronomers discovered that the universe is not only expanding but also accelerating. If you look at a distant galaxy and measure the velocity with which it is receding, that velocity is increasing with time. That would be extremely surprising if all the universe contained were ordinary matter and radiation, both of which have the gravitational effect of pulling things together and slowing down the expansion rate. A positive vacuum energy has the opposite effect: it pushes the universe apart, leading to accelerated expansion. Two teams of astronomers measured the distances and velocities of extragalactic supernovae, expecting to measure the deceleration of the universe. What they actually found was that it is speeding up. The discomfiting surprise at obtaining such an unexpected result was partly ameliorated by winning the Nobel Prize in 2011. (The debate "seems to" have been settled, because it's still an open possibility that cosmic acceleration is caused by something other than vacuum energy. But that's by far the leading explanation, on both theoretical and observational grounds.)

You might think that would be the end of it. Empty space has energy, we've measured it, cocoa and cupcakes all around.

But there's another question we're allowed to ask: What should we *expect* the vacuum energy to be? That's a funny question; since it's just a constant of nature, maybe we don't have the right to expect that it's any particular value at all. What we can do, however, is a quick-and-dirty estimate of how big we might guess the vacuum energy should be. The result is sobering.

The traditional way to estimate the vacuum energy is to distinguish between what the classical cosmological constant would be, and how quantum effects change that value. That's not really right; nature doesn't care that human beings like to start classically and build quantum mechanics on top of that. Nature is quantum from the start. But since all we're trying to do is get a very rough estimate, maybe this procedure is okay.

As it turns out, it's not okay. The quantum contribution to the vacuum energy is infinitely big. This kind of problem is endemic to quantum field theory; many calculations that we try to do by gradually including quantum effects end up giving us nonsensical, infinitely big answers.

But we shouldn't take those infinities too seriously. They can ultimately be traced to the fact that a quantum field can be thought of as a combination of vibrating modes at all different wavelengths, from incredibly long all the way down to zero. If we assume (for no especially good reason) that the classical minimum energy of each mode is zero, then the real-world vacuum energy is just the sum of all the additional quantum energies for each mode. Adding up the quantum energies for all those modes is what gives us an infinite vacuum energy. That's probably not physically realistic. After all, at very short distances we should expect spacetime itself to break down as a useful concept, as quantum gravity becomes impossible to ignore. It might make more sense to only include contributions with wavelengths larger than the Planck length, for example. We call this *imposing a cutoff*—looking at quantum field theory, but only including modes with wavelengths longer than a certain distance.

Unfortunately this doesn't quite fix the problem. If we estimate the quantum contribution to the vacuum energy by imposing a Planck-scale cutoff on the allowed modes, we get a finite answer rather than an infinite one, but that answer is 10^{122} times larger than the value we actually observe. This mismatch, known as the *cosmological constant problem*, has often been called the biggest discrepancy between theory and observation in all of physics.

The cosmological constant problem is not really a conflict between theory and observation in the strict sense. We don't have anything like a reliable theoretical prediction for what the vacuum energy should be.

Our very wrong estimate comes from making two dubious assumptions: that the classical contribution to the vacuum energy is zero, and that we impose a cutoff at the Planck scale. It's always possible that the classical contribution we should start with is almost exactly as large as the quantum piece, but with the opposite sign, so that when we add them together we get an observed "physical" vacuum energy with a relatively tiny value. We just have no idea why that should be true.

The problem is not that theory conflicts with observation; it's that our rough expectations are way off, which most people take as a clue that something mysterious and unknown is at work. Since the energy we estimated was a purely quantum-mechanical effect, and we measure its existence using its gravitational effect, it's plausible that we won't solve the problem until we have a fully working quantum theory of gravity.

∘ ∘ ∘

Popular discussions of quantum field theory will often describe the vacuum as full of "quantum fluctuations," or even "particles popping in and out of existence in empty space." That's an evocative picture, but it's more false than true.

In empty space described by the quantum-field-theory vacuum, nothing is fluctuating at all; the quantum state is absolutely stationary. The picture of particles popping in and out of existence is entirely different from the reality, in which the state is precisely the same from one moment to another. There is undoubtedly an intrinsically quantum contribution to the energy of empty space, but it's misleading to speak of that energy as coming from "fluctuations," when nothing is actually fluctuating. The system is sitting peacefully in its lowest-energy quantum state.

Why, then, are physicists constantly talking about quantum fluctuations? It's the same phenomenon we have noted in other contexts:

we human beings have an irresistible urge to think of *what we see* as being real, even though quantum mechanics keeps telling us to do better. Hidden-variable theories give in to this urge by making something real other than the smoothly evolving wave function.

Everettian quantum mechanics is clear: empty space is described by a stationary, unchanging quantum state, where nothing is happening from moment to moment. But if we were to look sufficiently carefully, measuring the values of a quantum field in some small region, we would see what looked like a random mess. And if we looked again a moment later, we would see a different-looking random mess. The temptation to conclude that there is something moving around in empty space, even when we're not looking, is overwhelming. But that's not what's going on. Rather, we're seeing a manifestation of what we talked about in the context of the uncertainty principle: when we observe a quantum state, we typically see something quite different from what the state was before we looked.

To drive this point home, imagine that we do a more experimentally feasible measurement. Rather than measuring the value of a field at every point, let's just measure the total number of particles in the vacuum state of a quantum field theory. In an ideal thought-experiment world, we can imagine doing that measurement throughout all of space all at once. Since by construction we're in the lowest-energy state, you won't be surprised to hear that we will, with perfect confidence, detect no particles anywhere. It's just empty space. But in the real world, we will be confined to doing an experiment in some specific region of space, such as the interior of our laboratory, and asking how many particles there are. What should we expect to see?

This doesn't sound like a hard question. If there are no particles anywhere, then certainly we won't see any particles in our lab, right? Alas, no. That's not how quantum field theory works. Even in the vacuum state, if our experimental probe is confined to some finite region, there will always be a small probability of observing one or more particles. Generally the probability will be really, really small—not something we have to worry about in realistic experimental setups—but it will be there. The converse is also true: there will be quantum states for which our local experiment will never see particles, but such states will have more energy overall than the vacuum state.

You might be tempted to ask: But are the particles *really there*? How can there be zero particles in the universe as a whole, and yet we might see particles when we look in any particular location?

But we're not dealing with a theory of particles; it's a theory of fields. Particles are what we see when we observe the theory in particular ways. We shouldn't be asking, "How many particles are there, really?" We should be asking, "What are the possible measurement outcomes when we observe a quantum state in this specific way?" A measurement of the form "How many particles are there in the entire universe?" is fundamentally different from one of the form "How many particles are there in this room?" So different that, just as for position and momentum, no quantum state will give definite answers for both questions at the same time. The number of particles we see isn't an absolute reality, it depends on how we look at the state.

∘ ∘ ∘

This leads us directly to an important property of quantum field theory: the entanglement between parts of the field in different regions of space.

Imagine dividing the universe into two regions by drawing an imaginary plane somewhere in space. Call the regions "left" and "right" for convenience. Classically, since fields live everywhere, to construct any particular field configuration we would have to specify what the field is doing both in the left region and in the right region. If there is a mismatch of the value of the fields across the boundary, that will correspond to a sharp discontinuity in the profile of the field overall. That's conceivable, but it costs energy for the field to change from point to point, so a discontinuous jump implies a large amount of energy at that point. This is why ordinary field configurations tend to vary smoothly, rather than suddenly.

At the quantum level, the classical statement "The field value tends to match across the boundary" turns into "The fields in the left and right regions tend to be highly entangled with each other." We can consider quantum states where the two regions are unentangled, but there would be an infinite amount of energy at the boundary.

This reasoning extends further. Imagine dividing up all of space into equal-sized boxes. Classically, the field would be doing something in each box, but to avoid infinite energy densities the values must match at the boundaries between boxes. In quantum field theory, therefore, what's happening in one box must be highly entangled with what's happening in neighboring boxes.

That's not all. If a box is entangled with its neighbors, and those neighboring boxes are entangled with their neighbors, it stands to reason that the fields in our original box should be entangled not only with its neighbors, but with the fields one box away. (That's not logically necessary, but it seems reasonable in this case, and a careful calculation affirms that it is true.) There will be a lot less entanglement with the fields one box away than for direct neighbors, but there will still be some there. And indeed this pattern continues all throughout space: the fields in any one box are entangled with the fields in every other box in the universe, although the amount of entanglement becomes less and less as we consider boxes that are farther and farther apart.

That may seem like a stretch, since after all there are an infinite number of boxes in an infinitely big universe. Can the fields in one little region, say, a single cubic centimeter, really be entangled with fields in every other cubic centimeter of the universe?

Yes, they can. In field theory, even a single cubic centimeter (or a box of any other size) contains an infinite number of degrees of freedom. Remember that we defined a *degree of freedom* in Chapter Four as

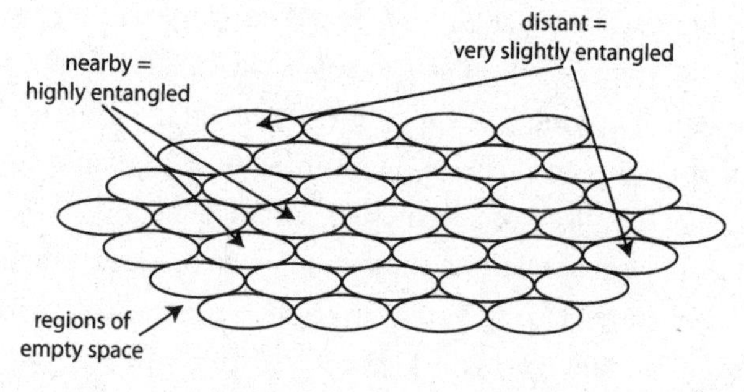

a number needed to specify the state of a system, such as "position" or "spin." In field theory, there are an infinite number of degrees of freedom in any finite region: at every point in space, the value of the field at that point is a separate degree of freedom. And there are an infinite number of points in space, even in just a small region.

Quantum-mechanically, the space of all the possible wave functions for a system is that system's Hilbert space. So the Hilbert space describing any region in quantum field theory is infinite-dimensional, because there are an infinite number of degrees of freedom. As we'll see, that might not continue to hold true in the correct theory of reality; there are reasons to think that quantum gravity features only a finite number of degrees of freedom in a region. But quantum field theory, without gravity, allows for infinite possibilities in any tiny box.

Those degrees of freedom share a lot of entanglement with the degrees of freedom elsewhere in space. To drive home just how much, imagine starting with the vacuum state, taking one of those one-cubic-centimeter boxes, and poking the quantum fields inside. By "poking" we mean any way we could conceivably imagine affecting the field just in that local region, by measuring it or otherwise interacting with it. We know that measuring a quantum state changes it into another state (indeed, to different states on each branch of the new wave function). Do you think that by poking the state strictly inside a given box, it's possible to instantly change the state outside the box?

If you know a little relativity, you might be tempted to answer "no"—it should take time for any effects to propagate to faraway regions. But then you remember the EPR thought experiment, where Alice's measurement on a spin can affect the quantum state of Bob's spin, no matter how far away they are from each other. Entanglement is the secret ingredient. And we just said that the vacuum state in quantum field theory is highly entangled, such that every box is entangled with every other box. Gradually you will begin to wonder whether poking the field in one box might be able to cause drastic changes in the rest of the state, even very far away.

Indeed it can. By poking a quantum field in one tiny region of space, it's possible to turn the quantum state of the whole universe into *literally any state at all*. Technically this result is known as the *Reeh-Schlieder theorem*, but it has also been called the *Taj Mahal theorem*. That's because it implies that without leaving my room, I can do an experiment and get an outcome that implies there is now, suddenly, a copy of the Taj Mahal on the moon. (Or any other building, at any other location in the universe.)

Don't get too excited. We can't purposefully force the Taj Mahal to be created, or reliably bring anything particular into existence. In the EPR example Alice can measure her spin, but she can't guarantee what outcome that measurement is going to get. The Reeh-Schlieder theorem implies that if we measure quantum fields locally, there is some measurement outcome we could get that would be associated with a Taj Mahal suddenly being on the moon. But no matter how hard we try, the probability of actually getting that outcome will be really, really, really tiny. Almost all the time, a local measurement leaves distant parts of the world pretty much unaltered. Like many remarkable results in quantum mechanics, it's not a practical worry.

A popular after-dinner discussion among certain circles is "Should we be surprised by the Reeh-Schlieder theorem, or not?" It certainly seems surprising that we can do a measurement in our basement that turns the state of the universe into literally anything. As surprising things go, that's up there. But the other side argues that once you understand entanglement, and appreciate that things can technically be possible but are so incredibly improbable that it really doesn't matter, we shouldn't be very surprised after all. Looked at in the right way, the potential for a Taj Mahal on the moon was there all along, in some tiny part of the quantum state. Our experiment simply lifted it out of the vacuum by branching the wave function in an appropriate way.

I think it's okay to be surprised. But more important, we should appreciate the richness and complexity of the vacuum. In quantum field theory, even empty space is an exciting place to be.

Breathing in Empty Space

Finding Gravity within Quantum Mechanics

Quantum field theory is able to successfully account for every experiment ever performed by human beings. When it comes to describing reality, it's the best approach we have. It's therefore extremely tempting to imagine that future physical theories will be set within the broad paradigm of quantum field theory, or perhaps small variations thereof.

But gravity, at least when it becomes strong, doesn't seem to be well described by quantum field theory. So in this chapter we'll ask whether we can make progress by attacking the problem from a different angle.

Following Feynman, physicists love to remind one another that nobody really understands quantum mechanics. Meanwhile, they have long lamented that nobody understands quantum gravity. Maybe these two lacks of understanding are related. Gravity, which describes the state of spacetime itself rather than just particles or fields moving within spacetime, presents special challenges when we try to describe it in quantum terms. Perhaps that shouldn't be surprising, if we don't think we fully understand quantum mechanics itself. It's possible that thinking about the foundations of quantum theory—in particular, the

Many-Worlds perspective that the world is just a wave function, and everything else emerges out of that—will shed new light on how curved spacetime emerges from quantum underpinnings.

Our self-appointed task is one of reverse engineering. Rather than taking classical general relativity and quantizing it, we will try to find gravity within quantum mechanics. That is, we will take the basic ingredients of quantum theory—wave functions, Schrödinger's equation, entanglement—and ask under what circumstances we can obtain emergent branches of the wave function that look like quantum fields propagating in a curved spacetime.

Up to this point in the book, basically everything we've talked about is either well understood and established doctrine (such as the essentials of quantum mechanics), or at least a plausible and respectable hypothesis (the Many-Worlds approach). Now we've reached the edge of what is safely understood, and will be venturing out into uncharted territory. We'll be looking at speculative ideas that might be important to understanding quantum spacetime and cosmology. But they might not be. Only years, possibly decades, of further investigation will reveal the answer with any confidence. By all means take these ideas as provocations to further thinking, and keep an eye on where the discussion goes in times to come, but keep in mind the intrinsic uncertainty that comes with wrestling with hard problems at the bleeding edge of our understanding.

◦ ◦ ◦

Albert Einstein once mused to a colleague, "On quantum theory I use more brain grease than relativity." But it was his contributions to relativity that made him an intellectual superstar.

Like "quantum mechanics," "relativity" does not refer to a specific physical theory, but rather a framework within which theories can be constructed. Theories that are "relativistic" share a common picture of the nature of space and time, one in which the physical world is described by events happening in a single unified "spacetime." Even before relativity, it was still possible to talk about spacetime in Newtonian physics: there is three-dimensional space, and one dimension of time, and to locate an event in the universe you have to specify both where the event is in space and when it occurs in time. But before Einstein, there wasn't much motivation for combining them into a single four-dimensional concept. Once relativity came along, that became a natural step.

There are two big ideas that go under the name of "the theory of relativity," the special theory and the general theory. *Special relativity*, which came together in 1905, is based on the idea that everyone measures light to travel at the same speed in empty space. Combining that insight with an insistence that there is no absolute frame of motion leads us directly to the idea that time and space are "relative." Spacetime is universal and agreed upon by everyone, but how we divvy it up into "space" and "time" will be different for different observers.

Special relativity is a framework that includes many specific physical theories, all of which are dubbed "relativistic." Classical electromagnetism, put together by James Clerk Maxwell in the 1860s, is a relativistic theory even though it was invented before relativity; the need to better understand the symmetries of electromagnetism was a driving force behind why relativity was invented in the first place. (Sometimes people misuse the word "classical" to include "non-relativistic," but it's better to reserve it to mean "non-quantum.") Quantum mechanics and special relativity are 100 percent compatible with each other. The quantum field theories used in modern particle physics are relativistic to their cores.

The other big idea in relativity came ten years later, when Einstein proposed *general relativity*, his theory of gravity and curved spacetime. The crucial insight was that four-dimensional spacetime isn't just a static background on which the interesting parts of physics take place; it has a life of its own. Spacetime can bend and warp, and does so in response to the presence of matter and energy. We grow up learning about the flat geometry described by Euclid, in which initially parallel lines remain parallel forever and the angles inside a triangle always add up to 180 degrees. Spacetime, Einstein realized, has a non-Euclidean geometry, in which these venerable facts are no longer the case. Initially parallel rays of light, for example, can be focused together while moving through empty space. The effects of this warping of geometry are what we recognize as "gravity." General relativity came with numerous mind-stretching consequences, such as the expansion of the universe and the existence of black holes, though it has taken physicists a long time to appreciate what those consequences are.

Special relativity is a framework, but general relativity is a specific theory. Just like Newton's laws govern the evolution of a classical system or the Schrödinger equation governs the evolution of a quantum wave function, Einstein derived an equation that governs the curvature of spacetime. As with Schrödinger's equation, it's fun to actually see Einstein's equation written out, even if we don't bother with all the details:

$$R_{\mu\nu} - (1/2) R g_{\mu\nu} = 8\pi G T_{\mu\nu}$$

The math behind Einstein's equation is formidable, but the basic idea is simple, and was pithily summarized by John Wheeler: matter tells spacetime how to curve, and spacetime tells matter how to move. The left-hand side measures the curvature of spacetime, while the right-hand side characterizes energy-like quantities, including momentum, pressure, and mass.

General relativity is classical. The geometry of spacetime is unique, evolves deterministically, and can in principle be measured to arbitrary precision without disturbing it. Once quantum mechanics came along, it was perfectly natural to try to "quantize" general relativity, obtaining a quantum theory of gravity. Easier said than done. What makes relativity special is that it's a theory of spacetime rather than a theory of stuff within spacetime. Other quantum theories describe wave functions that assign probabilities to observing things at definite, well-defined locations in space and moments in time. Quantum gravity, by contrast, will have to be a quantum theory of spacetime itself. That raises some issues.

Einstein, naturally, was one of the first to appreciate the problem. In 1936, he mused on the difficulty of even imagining how to apply the principles of quantum mechanics to the nature of spacetime:

> Perhaps the success of the Heisenberg method points to a purely algebraical method of description of nature, that is to the elimination of continuous functions from physics. Then, however, we must also give up, by principle, the space-time continuum. It is not unimaginable that human ingenuity will some day find methods which will make it possible to proceed along such a path. At the present time, however, such a program looks like an attempt to breathe in empty space.

Here Einstein is contemplating Heisenberg's approach to quantum theory, which you'll remember provided a description in terms of explicit quantum jumps without trying to fill in the details about microscopic processes happening along the way. Similar worries persist if we switch to a more Schrödingerian point of view with wave functions. Presumably we would need a wave function that assigns amplitudes to different possible geometries of spacetime. But if we imagine, for example, two branches of such a wave function that describe different spacetime geometries, there is no unique way of specifying that two events in the two branches correspond to the "same" point in spacetime. There is no unique map, in other words, between two different geometries.

Consider a two-dimensional sphere and torus. Imagine that a friend of yours picks out a point on a sphere, and then asks you to pick out "the same" point on the torus. You'd be stymied, and for good reason; there's no way to do it.

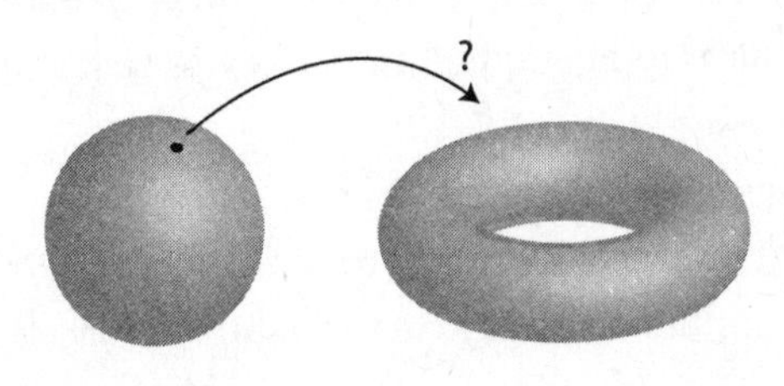

Apparently, spacetime can't play the same central role in quantum gravity that it does in the rest of physics. There isn't a single spacetime, there's a superposition of many different spacetime geometries. We can't ask what the probability might be to find an electron at a certain point in space, since there's no objective way to specify which point we're talking about.

Quantum gravity, then, comes with a set of conceptual issues that distinguish it from other quantum-mechanical theories. These issues can have important ramifications for the nature of our universe, including the question of what happened at the beginning, or if there was a beginning at all. We can even ask whether space and time are themselves fundamental, or if they emerge out of something deeper.

° ° °

Just like the foundations of quantum mechanics, the field of quantum gravity was relatively ignored for decades as physicists concentrated on other things. Not completely; Hugh Everett was inspired to propose the Many-Worlds approach in part by thinking about the quantum theory of the entire universe, where gravity plays an important role, and his mentor, John Wheeler, worried about the problem for years. But even putting aside the conceptual issues, other obstacles got in the way of making serious progress on quantizing gravity.

A major roadblock is the difficulty of getting direct experimental data. Gravity is a very weak force; the electric repulsion between two electrons is about 10^{43} times stronger than their gravitational attraction. In any realistic experiment involving just a few particles, where we might expect quantum effects to be visible, the force of gravity is utterly negligible compared to other influences. We can imagine building a particle accelerator powerful enough to smash particles together at the Planck energy, where quantum gravity should become important. Unfortunately, if we simply scale up the technology in current machines, the resulting accelerator would have to be light-years in diameter. It's not a feasible construction project at this time.

There are also technical problems with the theory itself, in addition to the conceptual ones just mentioned. General relativity is a classical field theory. The field involved is called the *metric*. (The symbol $g_{\mu\nu}$ in the middle of Einstein's equation represents the metric, and the other quantities depend on it.) The word "metric" ultimately derives from the Greek *metron*, "something used to measure," and that's exactly what the metric field allows us to do. Given a path through spacetime, the metric tells us the distance along that path. The metric essentially updates Pythagoras's theorem, which works in flat Euclidean geometry but has to be generalized when spacetime is curved. Knowing the length of every curve suffices to fix the geometry of spacetime at every point.

Spacetime has a metric even in special relativity, or for that matter in Newtonian physics. But that metric is rigid, unchanging, and flat—the curvature of spacetime is zero at every point. The big insight of general relativity was to make the metric field into something that is dynamical and affected by matter and energy. We can attempt to quantize that field just as we would any other. Small ripples in the quantized gravitational field look like particles called *gravitons*, just like ripples in the electromagnetic field look like photons. Nobody has ever detected a graviton, and it's possible that nobody ever will, since the gravitational force is so incredibly weak. But if we accept the basic principles of general relativity and quantum mechanics, the existence of gravitons is inevitable.

We can then ask what happens when gravitons scatter off each other or off other particles. Sadly, what we find is that the theory predicts nonsense, if it predicts anything at all. An infinite number of input parameters are needed to calculate any particular quantity of interest, so the theory has no predictive power. We can restrict our attention to an "effective" field theory of gravity, where by fiat we limit our attention to long wavelengths and low energies. That's what allows us to calculate the gravitational field in the solar system, even in quantum gravity. But if we want a theory of everything, or at least a theory of gravity that is valid at all possible energies, we're stuck. Something dramatic is called for.

The most popular contemporary approach to quantum gravity is *string theory*, which replaces particles by little loops or segments of one-dimensional "string." (Don't ask what the strings are made of—string stuff is what everything else is made of.) The strings themselves are incredibly small, so much so that they appear like particles when we observe them from a distance.

String theory was initially proposed to help understand the strong nuclear force, but that didn't work out. One of the problems was that the theory inevitably predicts the existence of particles that look and behave exactly like gravitons. That was initially perceived as an annoyance, but pretty soon physicists thought to themselves, "Hmm, gravity actually exists. Maybe string theory is a quantum theory of gravity?" That turns out to be true, and even better there is a bonus: the theory makes finite predictions for all physical quantities, without needing an infinite number of input parameters. The popularity of strings exploded in 1984 when Michael Green and John Schwarz showed that the theory is mathematically consistent.

Today string theory is the most pursued approach to exploring quantum gravity by a wide margin, although other ideas maintain their adherents. The second-most-popular approach is *loop quantum gravity*, which began as a way of directly quantizing general relativity by using a clever choice of variables—rather than looking at the curvature of spacetime at each point, we consider how vectors are rotated when they travel around closed loops in space. (If space is flat, they don't rotate at all, while if space is curved, they can rotate by a lot.) String theory aspires to be a theory of all the forces and matter at once, while loop quantum gravity only aims at gravity itself. Unfortunately, the obstacles to gathering experimental data relevant to quantum gravity are equally formidable for all the alternatives, so we're stuck not really knowing which approach (if any) is on the right track.

While string theory has been somewhat successful in dealing with the technical problems of quantum gravity, it hasn't shed much light on the conceptual problems. Indeed, one way of thinking about different approaches within the quantum-gravity community is to ask how we should think about the conceptual side of things. A string theorist is likely to believe that if we take care of all the technical issues, the conceptual problems will eventually resolve themselves. Someone who thinks otherwise might be nudged toward loop quantum gravity or another alternative approach. When the data don't point one way or the other, opinions tend to become deeply entrenched.

String theory, loop quantum gravity, and other ideas share a common pattern: they start with a set of classical variables, then quantize. From the perspective we've been following in this book, that's a little backward. Nature is quantum from the start, described by a wave function evolving according to an appropriate version of the Schrödinger equation. Things like "space" and "fields" and "particles" are useful ways of talking about that wave function in an appropriate classical limit. We don't want to start with space and fields and quantize them; we want to extract them from an intrinsically quantum wave function.

° ° °

How can we find "space" within a wave function? We want to identify features of the wave function that resemble space as we know it, and in particular something that would correspond to a metric that defines distances. So let's think about how distances show up in ordinary quantum field theory. For simplicity, let's just think about distances in space; we'll talk later about how time might enter into the game.

There's one obvious place that distances show up in quantum field theory, which we've seen in the last chapter: in empty space, fields in different regions are entangled with each other, and regions that are far away are less entangled than ones that are nearby. Unlike "space," the concept of "entanglement" is always available to us in any abstract quantum wave function. So perhaps we can get some purchase here, looking at the entanglement structure of states and using that to define distances. What we need is a quantitative measure of how entangled a quantum subsystem actually is. Happily, such a measure exists: it's the entropy.

John von Neumann showed how quantum mechanics introduces a notion of entropy that parallels the classical definition. As explained by Ludwig Boltzmann, we start with a set of constituents that can mix together in various ways, like atoms and molecules in a fluid. The entropy is then a way of counting the number of ways those constituents can be arranged without changing the macroscopic appearance of the system. Entropy is related to ignorance: high-entropy states are those for which we don't know much about the microscopic details of a system just from knowing its observable features.

Von Neumann entropy, meanwhile, is purely quantum mechanical in nature, and arises from entanglement. Consider a quantum system that is divided into two parts. It could be two electrons, or the quantum fields in two different regions of space. The system as a whole is described by a wave function, as usual. It has some definite quantum state, even if we can only predict measurement outcomes probabilistically. But as long as the two parts are entangled, there is only the one wave function for the whole thing, not a separate wave function for each part. The parts, in other words, are not in definite quantum states of their own.

Von Neumann showed that, for many purposes, the fact that entangled subsystems don't have definite wave functions of their own is analogous to having a wave function, but we just don't know what it is. Quantum subsystems, in other words, closely resemble the classical situation where there are many possible states that look macroscopically the same. And this uncertainty can be quantified into what we now call the *entanglement entropy*. The higher the entropy of a quantum subsystem, the more it's entangled with the outside world.

Think about two qubits, one belonging to Alice and the other to Bob. It might be that they are unentangled, so each qubit has its own wave function, for example, an equal superposition of spin-up and spin-down. In that case, the entanglement entropy of each qubit is zero. Even if we can only predict measurement outcomes probabilistically, each subsystem is still in a definite quantum state.

But imagine that the two qubits are entangled, in an equal superposition of "both qubits are spin-up" and "both qubits are spin-down." Alice's qubit doesn't have its own wave function, because it's entangled with Bob's. Indeed, Bob could perform a measurement of his spin, branching the wave function, so that now there are two copies of Alice, each of whom has a spin in a definite state. But neither copy of Alice knows which state that is; she's in a state of ignorance, where the best she can do is say that there is a fifty-fifty chance her qubit is spin-up or spin-down. Note the subtle difference: Alice's qubit is not in a quantum superposition where she doesn't know what the measurement outcome will be; it's in a state on each branch that will give a definite measurement outcome, but she doesn't know which state it is. We therefore describe her qubit as having a nonzero entropy. Von Neumann's idea was that we should ascribe a nonzero entropy to Alice's qubit even before Bob measures his, because after all she doesn't even know whether he's done a measurement. That's the entanglement entropy.

∘ ∘ ∘

Let's see how entanglement entropy appears in quantum field theory. Forgetting about gravity for a second, consider a region of empty space in the vacuum state, specified by a boundary separating inside the region from outside. Empty space is a richly textured place, full of quantum degrees of freedom that we can think of as modes of vibrating fields. The modes inside the region will be entangled with the modes outside, so the region has an entropy associated with it, even if the overall state is simply the vacuum.

We can even calculate what that entropy is. The answer is: infinity. This is a common complication with quantum field theory, that many questions of apparent physical relevance have seemingly infinite answers because there are an infinite number of possible ways for a field to vibrate. But just as we did for the vacuum energy in the last chapter, we can ask what happens when we impose a cutoff, allowing only modes longer than a certain wavelength. The resulting entropy is finite, and it turns out to be naturally proportional to the area of the region's boundary. The reason isn't hard to understand: field vibrations in one part of space are entangled with regions all over, but most of the entanglement is concentrated on nearby regions. The total entropy of a region of empty space depends on the amount of entanglement across the boundary, which is proportional to how big that boundary is—its area.

That's an intriguing feature of quantum field theory. Pick out a region within empty space, and the entropy of that region is proportional to the area of its boundary. That relates on the one hand a geometric quantity, the area of a region, to a "matter" quantity, the entropy contained inside. It all sounds vaguely reminiscent of Einstein's equation,

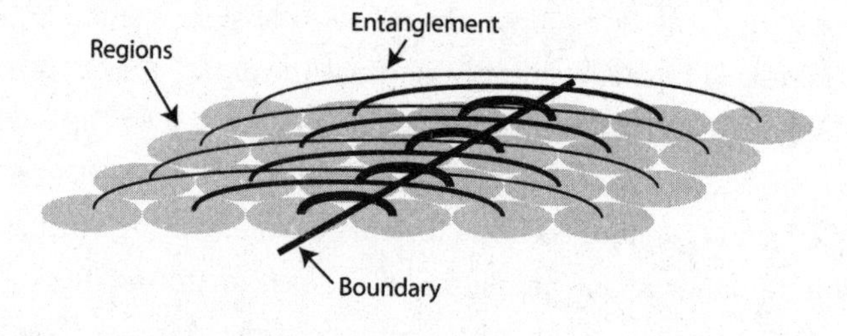

which also connects geometry (the curvature of spacetime) to a matter quantity (energy). Are they somehow related?

They could be, as was pointed out in a provocative 1995 paper by Ted Jacobson, an ingenious physicist at the University of Maryland. In ordinary quantum field theory without gravity, entropy is proportional to area in the vacuum state, but in higher-energy states it doesn't have to be. Jacobson postulated that there's something special about gravity: when gravity is included, the entropy of a region is *always* proportional to its boundary area. That's not at all what we would expect in quantum field theory, but maybe it happens once gravity enters the game. We can imagine that it might be the case, and see what happens.

What happens is pretty wonderful. Jacobson posited that the area of a surface is proportional to the entropy of the region it encloses. Area is a geometric quantity; we can't calculate the area of a surface without knowing something about the geometry of the space it is a part of. Jacobson noted that we could relate the area of a very small surface to the same geometric quantity that appears on the left-hand side of Einstein's equation. Meanwhile, entropy tells us something about "matter," broadly construed; about the stuff that is living within spacetime. The concept of entropy originally arose within thermodynamics, where it was related to the heat leaving a system. And heat is a form of energy. Jacobson also argued that this entropy could be directly related to the energy term appearing on the right-hand side of Einstein's equation. Through these maneuvers he was able to *derive* Einstein's equation for general relativity, rather than directly postulating it, as Einstein did.

To say the same thing more directly, we consider a small region in flat spacetime. It has some entropy, because the modes inside the region are entangled with those outside. Now imagine changing the quantum state a little bit, so that we decrease the amount by which that region is entangled, and therefore decrease its entropy. In Jacobson's picture, the area bounding our region changes in response, shrinking by a bit. And he shows that this response of the geometry of spacetime to a change in the quantum state is equivalent to Einstein's equation of general relativity, relating curvature to energy.

This was the beginning of a surge of interest in what is now called "entropic" or "thermodynamic" gravity; other important contributions were made by Thanu Padmanabhan (2009) and Erik Verlinde (2010). The behavior of spacetime in general relativity can be thought of as simply the natural tendency of systems to move toward configurations of higher entropy.

This is a fairly radical change of perspective. Einstein thought in terms of energy, a definite quantity associated with particular configurations of stuff in the universe. Jacobson and others have argued that we can reach the same conclusions by thinking about entropy, a collective phenomenon that emerges from the mutual interaction of many small constituents of a system. This simple shift in focus might offer a crucial way forward in our quest to discover a fundamentally quantum theory of gravity.

○ ○ ○

Jacobson wasn't himself proposing a theory of quantum gravity; he was pointing to a new way to derive Einstein's equation for classical general relativity, with quantum fields acting as the source of energy. The appearance of words like "area" and "region of space" should indicate to us that the above discussion treated spacetime as a tangible, classical thing. But given the central role that entanglement entropy plays in his derivation, it's natural to ask whether we might adapt the basic ideas to an approach that is more intrinsically quantum from the start, where space itself emerges from the wave function.

In Many-Worlds, a wave function is just an abstract vector living within the super-high-dimensional mathematical construct of Hilbert space. Usually we make wave functions by starting with something classical and quantizing it, which gives us an immediate handle on what the wave function is supposed to represent, the basic parts from which it is constructed. But here we don't have any such luxury. All we have is the state itself and Schrödinger's equation. We speak abstractly of "degrees of freedom," but they aren't the quantized version of any readily identifiable classical stuff—they are the quantum-mechanical essence out of which spacetime, and everything else, emerges. John Wheeler used to talk about the idea of "It from Bit," suggesting that the physical world arose (somehow) out of information. These days, when entanglement of quantum degrees of freedom is the main focus, we like to talk about "It from Qubit."

If we look back at the Schrödinger equation, it says that the rate at which the wave function changes with time is governed by the Hamiltonian. Remember that the Hamiltonian is a way of describing how much energy the system contains, and it's a compact way of capturing all of the system's dynamics. A standard feature of Hamiltonians in the real world is dynamical locality—subsystems interact with other subsystems only when they are next to each other, not when they are far away. Influences can travel through space, but only at speeds less than or equal to the speed of light. So an event at one particular moment only immediately affects what's going on at its present location.

With the problem we've assigned to ourselves—how does space emerge from an abstract quantum wave function?—we don't have the convenience of starting with individual parts and asking how they interact. We know what "time" means in this context—it's right there in the Schrödinger equation, the letter t—but we don't have particles, or fields, or even locations in a three-dimensional world. We're caught breathing in empty space, and need to look for oxygen where we can find it.

Happily, this is a case where reverse engineering works quite well. Rather than starting with individual pieces of a system and asking how they interact, we can go the other way around: Given the system as a whole (the abstract quantum wave function) and its Hamiltonian, is there a sensible way to break it up into subsystems? It's like buying sliced bread all your life, and then being handed an un-sliced loaf. There are many ways we could imagine slicing it; is there one particular way that's clearly the best?

Yes, there is, if we believe that locality is an important feature of the real world. We can tackle the problem bit by bit, or qubit by qubit, at any rate.

A generic quantum state can be thought of as a superposition of a set of basis states with definite fixed energy. (Just like a generic state of a spinning electron can be thought of as a superposition of an electron that is definitely spin-up and one that is definitely spin-down.) The Hamiltonian tells us what the actual energy is for each possible definite-energy state. Given that list of possible energies, we can ask whether any particular way of dividing the wave function into subsystems implies that those subsystems interact "locally." In fact, for a random list of energies, there won't be any way of dividing the wave function into local subsystems, but for the right kind of Hamiltonian, there will be exactly one such way. Demanding that physics look local tells us how to decompose our quantum system into a collection of degrees of freedom.

In other words, we don't need to start with a set of fundamental building blocks of reality, then stick them together to make the world. We can start with the world, and ask if there is a way to think about it as a collection of fundamental building blocks. With the right kind of Hamiltonian, there will be, and all of our data and experience of the world suggests that we do have the right kind of Hamiltonian. It's easy to imagine possible worlds where the laws of physics weren't local at all.

But it's hard to imagine what life would be like in such a world, or even whether life would be possible; the locality of physical interactions helps bring order to the universe.

∘ ∘ ∘

We can begin to see how space itself emerges from the wave function. When we say that there's a unique way of dividing up our system into degrees of freedom that interact locally with their neighbors, all we really mean is that each degree of freedom interacts with only a small number of other degrees of freedom. The notions of "local" and "nearest" aren't imposed from the start—they pop out from the fact that these interactions are very special. The way to think about it isn't "degrees of freedom interact only when they are nearby," but rather "we *define* two degrees of freedom to be 'nearby' when they directly interact with each other, and 'far away' when they don't." A long list of abstract degrees of freedom has been knit together into a network, in which each degree of freedom is connected to a small number of other ones. This network forms the skeleton on which space itself is constructed.

That's a start, but we want to do even more. When someone asks you how far apart two different cities are, they're looking for something a bit more specific than "near" or "far." They want an actual distance, and that's what the metric on spacetime ordinarily lets us calculate. In our abstract wave function divided up into degrees of freedom, we haven't yet constructed a full geometry, just a notion of near and far.

We can do better. Remember the intuition from vacuum states in quantum field theory that Jacobson used to derive Einstein's equation: the entanglement entropy of a region of space is proportional to the area of its boundary. In our current context of a quantum state described in terms of abstract degrees of freedom, we don't know what "area" is supposed to mean. But we do have entanglement between the degrees of freedom, and for any collection of them we can compute their entropy.

So once again following our reverse-engineering philosophy, we can *define* the "area" of a collection of degrees of freedom to be proportional to its entanglement entropy. In fact, we can assert this for every possible subset of degrees of freedom, assigning areas to every surface we can imagine drawing within our network. Happily, mathematicians long ago figured out that knowing the area of every possible surface in a region is enough to fully determine the geometry of that region; it's completely equivalent to knowing the metric everywhere. In other words, the combination of (1) knowing how our degrees of freedom are entangled, and (2) postulating that the entropy of any collection of degrees of freedom defines an area of the boundary around that collection, suffices to fully determine the geometry of our emergent space.

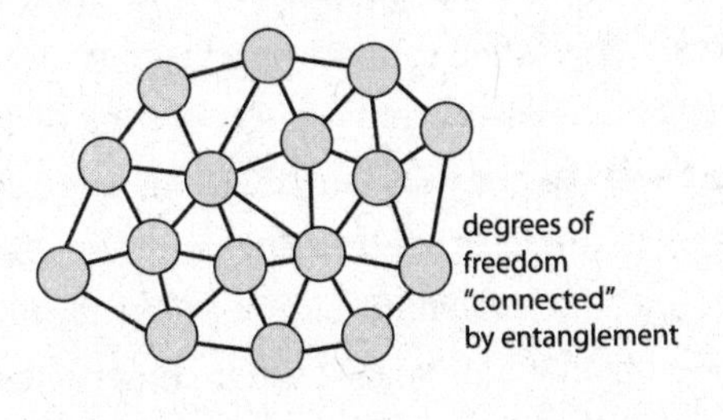

We can describe this construction in equivalent but slightly less formal terms. Pick out two of our spacetime degrees of freedom. They will generally have some entanglement between them. If they were modes of vibrating quantum fields in the vacuum state, we know exactly what that degree of entanglement would be: it would be high if they were nearby, and low if they were far away. Now we are simply thinking the other way around. If the degrees of freedom are highly entangled, we *define* them to be nearby, and the farther and farther away, the less entangled they are. A metric on space has emerged from the entanglement structure of the quantum state.

Thinking this way is a bit unusual, even for physicists, because we're used to thinking of particles moving through space, while taking

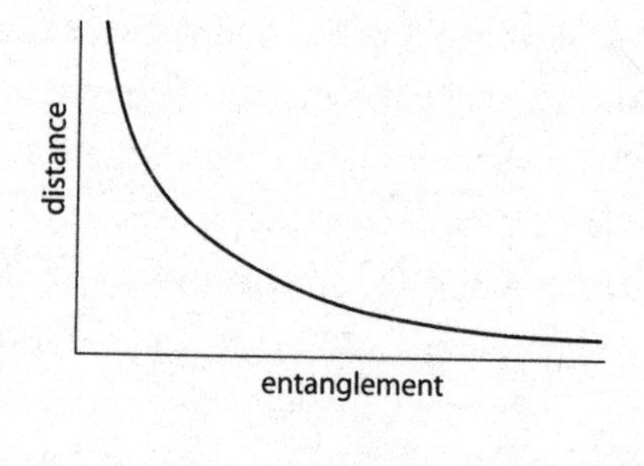

space itself for granted. As we know from the EPR thought experiment, two particles can be completely entangled no matter how far away they are; there's no necessary relationship between entanglement and distance. Here, however, we're not talking about particles but about the fundamental building-block degrees of freedom that make up space itself. Those aren't entangled in any old way; they are strung together in a very specific structure.*

Now we can use Jacobson's trick with entropy and area. Knowing the area of every surface in our network gives us a geometry, and knowing the entropy of each region tells us something about the energy in that region. I've been involved with this approach myself, in papers from 2016 and 2018 with my collaborators ChunJun (Charles) Cao and Spyridon Michalakis. Closely related ideas have been investigated by Tom Banks, Willy Fischler, Steve Giddings, and other physicists who are willing to contemplate the idea that spacetime isn't fundamental, but emerges from the wave function.

We aren't quite at the point where we can simply say, "Yes, this emergent geometry on space evolves with time in exactly the right way

* In 2013, Juan Maldacena and Leonard Susskind suggested that we should think of entangled particles as being connected by a microscopic (and impossible-to-travel-through) wormhole in spacetime. This has been dubbed the "ER=EPR conjecture," after two famous papers from 1935: one by Einstein and Nathan Rosen, where they introduced the concept of wormholes; and the other of course by Einstein, Rosen, and Boris Podolsky, where they discussed entanglement. How far such a suggestion can be taken is still unclear.

to describe a spacetime that obeys Einstein's equation of general relativity." That's the ultimate goal, but we're not there yet. What we can do is to specify a list of requirements under which that's exactly what does happen. The individual requirements seem reasonable—things like "at long distances, physics looks like an effective quantum field theory"—but many of them remain unproven as yet, and so far the most rigorous results are available only in situations where the gravitational field is relatively weak. We don't yet have a way of describing black holes or the Big Bang, though there are some promising ideas.

That's life as a theoretical physicist. We don't have all the answers, but let's not lose sight of the overall ambition: starting from an abstract quantum wave function, we have a road map describing how space emerges, with a geometry fixed by quantum entanglement, and that geometry seems to obey the dynamical rules of general relativity. There are so many caveats and assumptions going into this proposal that it's hard to know where to start listing them. But there seems to be a very real prospect that the route to understanding the universe lies not in quantizing gravity, but in finding gravity within quantum mechanics.

○ ○ ○

You may have noticed a tiny imbalance in this discussion. We've been asking how spacetime can emerge from entanglement in quantum gravity. But if we're honest, we've really only looked at how space emerges; we've taken time for granted as something that comes along for the ride. And it's possible that this approach is completely fair. Although relativity treats space and time as if they were on an equal footing, quantum mechanics generally does not. The Schrödinger equation, in particular, treats them very differently: it literally describes how the quantum state evolves with time. "Space" may or may not be part of that equation, depending on what system we're looking at, but time is fundamental. It's plausible that the symmetry between space and time that we're familiar with from relativity isn't built into quantum gravity, but emerges in the classical approximation.

It is nevertheless overwhelmingly tempting to wonder whether time, like space, might be emergent rather than fundamental, and whether entanglement might have anything to do with it. The answer is yes on both counts, although the details remain a little sketchy.

If we take the Schrödinger equation at face value, time seems to be right there in a fundamental way. Indeed, it immediately follows that the universe lasts eternally toward both the past and future, for almost all quantum states. You might think that this conflicts with the oft-repeated fact that the Big Bang was the beginning of our universe, but we don't actually know that oft-repeated fact to be true. That's a prediction of classical general relativity, not of quantum gravity. If quantum gravity operates according to some version of the Schrödinger equation, then for almost all quantum states, time runs from minus infinity in the past to plus infinity in the future. The Big Bang might be simply a transitional phase, with an infinitely old universe preceding it.

We have to say "almost all" in these statements because there is one loophole. The Schrödinger equation says that the rate of change of the wave function is driven by how much energy the quantum system has. What if we consider systems whose energy is precisely zero? Then all the equation says is that the system doesn't evolve at all; time has disappeared from the story.

You might think it's extremely implausible that the universe has exactly zero energy, but general relativity suggests you shouldn't be so sure. Of course there seem to be energy-containing things all around us—stars, planets, interstellar radiation, dark matter, dark energy, and so on. But when you go through the math, there is also a contribution to the energy of the universe from the gravitational field itself, which is generally negative. In a closed universe—one that wraps around on itself to form a compact geometry, like a three-dimensional sphere or torus, rather than stretching to infinity—that gravitational energy precisely cancels the positive energy from everything else. A closed universe has exactly zero energy, regardless of what's inside.

That's a classical statement, but there's a quantum-mechanical analogue that was developed by John Wheeler and Bryce DeWitt. The Wheeler-DeWitt equation simply says that the quantum state of the universe doesn't evolve at all as a function of time.

This seems crazy, or at least in flagrant contradiction to our observational experience. The universe certainly seems to evolve. This puzzle has been cleverly labeled the *problem of time* in quantum gravity, and it is where the possibility of emergent time might come to the rescue. If the quantum state of the universe obeys the Wheeler-DeWitt equation (which is plausible, but far from certain), time has to be emergent rather than fundamental.

One way that might work was suggested by Don Page and William Wootters in 1983. Imagine a quantum system consisting of two parts: a clock, and everything else in the universe. Imagine that both the clock and the rest of the system evolve in time as usual. Now take snapshots of the quantum state at regular intervals, perhaps once per second or once per Planck time. In any particular snapshot, the quantum state describes the clock reading some particular time, and the rest of the system in whatever configuration it was in at that time. That gives us a collection of instantaneous quantum states of the system.

The great thing about quantum states is that we can simply add them together (superposing them) to make a new state. So let's make a new quantum state by adding together all of our snapshots. This new quantum state doesn't evolve over time; it just exists, as we constructed it by hand. And there is no specific time reading on the clock; the clock subsystem is in a superposition of all the times at which we took snapshots. It doesn't sound much like our world.

But here's the thing: within that superposition of all the snapshots, the state of the clock is entangled with the state of the rest of the system. If we measure the clock and see that it reads some particular time, then the rest of the universe is in whatever state our original evolving system was caught in at precisely that time.

$$\begin{aligned} \Psi = &(\text{system @ t=0, clock = 0}) \\ &+ (\text{system @ t=1, clock = 1}) \\ &+ (\text{system @ t=2, clock = 2}) \\ &+ \ldots \end{aligned}$$

In other words, there's not "really" time in the superposition state, which is completely static. But entanglement generates a relationship between what the clock reads and what the rest of the universe is doing. And the state of the rest of the universe is precisely what it would be if it were evolving as the original state did over time. We have replaced "time" as a fundamental notion with "what the clock reads in this part of the overall quantum superposition." In that way, time has emerged from a static state, thanks to the magic of entanglement.

The jury remains out on whether the energy of the universe actually is zero, and therefore time is emergent, or it is any other number, such that time is fundamental. At the current state of the art, it makes sense to keep our options open and investigate both possibilities.

14

Beyond Space and Time

Holography, Black Holes, and the Limits of Locality

Before Stephen Hawking's death in 2018, he was the most famous living scientist in the world by a comfortable margin. That notoriety was entirely deserved; not only was Hawking a charismatic and influential public figure, and not only did he have an inspirational personal story, but his scientific contributions were incredibly significant in their own right.

Hawking's greatest achievement was showing that, once we include the effects of quantum mechanics, black holes "ain't so black," as he liked to say. Black holes actually emit a steady stream of particles out into space, and those particles carry energy away from the black hole, causing it to shrink in size. This realization led both to profound insights (black holes have entropy) as well as unexpected puzzles (where does the information go when black holes form and then evaporate away?).

The fact that black holes radiate, and the implications of that surprising idea, are the single best clue we have about the nature of quantum gravity. Hawking didn't first construct a full theory of quantum gravity and then use it to show that black holes radiate. Instead, he used a reasonable approximation, treating spacetime itself as classical, with dynamical quantum fields living on top of it. We hope that this is a reasonable approximation, anyway; but some of the puzzling aspects of Hawking's insight have given us second thoughts. Forty-five years after Hawking's original paper on the subject, trying to understand black-hole radiation is still one of the hottest topics in contemporary theoretical physics.

While that task is far from complete, one implication seems clear: the simple picture sketched in the last chapter, where space emerges from a set of entangled nearest-neighbor degrees of freedom, is probably not the entire story. It's a very good story, and might be the right starting point for constructing a theory of quantum gravity. But it relies heavily on the idea of locality—what happens at one point in space can have an immediate effect only on points right next door. Black holes, to the extent that we understand them, seem to be indicating that nature is more subtle than that. In some circumstances the world looks like a collection of degrees of freedom interacting with their nearest neighbors, but when gravity becomes strong, that simple picture breaks down. Rather than being distributed throughout space, degrees of freedom squeeze together on a surface, and "space" is merely a holographic projection of the information contained therein.

Locality undoubtedly plays an important role in our everyday lives, but it seems like the fundamental nature of reality can't quite be captured by a set of things happening at precise locations in space. Once again, what we have here is a job for the Many-Worlds approach to quantum mechanics. Other formulations take space as a given and work within it; the wave-function-first Everettian philosophy allows us to accept that space can appear fundamentally different depending on how we look at it, if it's a useful concept at all. Physicists are still wrestling with the implications of this idea, but it's already led us to some very interesting places indeed.

° ° °

In general relativity, a black hole is a region of spacetime that is curved so dramatically that nothing can escape from it, not even light itself. The edge of the black hole, demarcating the inside from the outside, is the *event horizon*. According to classical relativity, the area of the event horizon can only grow, not shrink; black holes increase in size when matter and energy fall in, but cannot lose mass to the outside world.

Everyone thought that was true in nature until 1974, when Hawking announced that quantum mechanics changes everything. In the presence of quantum fields, black holes naturally radiate particles into their surroundings. Those particles have a blackbody spectrum, so every black hole has a temperature; more massive black holes are cooler, while very small black holes are incredibly hot. The formula for the temperature of a black hole's radiation is engraved on Hawking's gravestone in Westminster Abbey.

Particles radiated by a black hole carry away energy, causing the hole to lose mass and eventually evaporate away completely. While it would be nice to observe Hawking radiation in a telescope, it's not going to happen for any of the black holes we know about. The Hawking temperature of a black hole the mass of the sun would be about 0.00000006 Kelvin. Any such signal would be swamped by other sources, such as the leftover microwave radiation from the Big Bang, which has a temperature of about 2.7 Kelvin. Even if such a black hole never grew by accreting matter and radiation, it will take over 10^{67} years for it to evaporate away completely.

There is a standard story that is told to explain why black holes emit radiation. I've told it, Hawking has told it, everyone tells it. It goes like this: according to quantum field theory, the vacuum is a bubbling stew of particles popping in and out of existence, typically in pairs consisting of one particle and one anti-particle. Ordinarily we don't notice, but in the vicinity of a black hole event horizon, one of the particles can fall inside the hole and then never get out, while the other escapes to the outside world. From the perspective of someone watching from afar, the escaping particle has positive energy, so to balance the books the infalling particle must have negative energy, and the black hole shrinks in mass as it absorbs these negative-energy particles.

Given our wave-functions-first Everettian perspective, there's a more accurate way to describe what's happening. The particles-appearing-and-disappearing story is a colorful metaphor that often provides physical intuition, and this is definitely one of those cases. But what we really have is a quantum wave function of the fields near the black hole. And that wave function is not static; it evolves into something else, in this case a smaller black hole plus some particles traveling away from it in all directions. It's not that different from an atom whose electrons have a bit of extra energy, and which therefore drop down to lower-energy states by emitting photons. The difference is that the atom eventually reaches a state of lowest possible energy and stays there, while the black hole (as far as we understand) just decays away entirely, exploding at the last second in a flash of high-energy particles.

The story of how black holes radiate and evaporate was derived by Hawking using the techniques of conventional quantum field theory, just in a curved spacetime of general relativity rather than a particle physicist's usual no-gravity context. It's not a genuinely quantum-gravity result; spacetime itself is treated classically, not as part of the quantum wave function. But nothing about the scenario actually seems to require deep knowledge of quantum gravity. As far as physicists can tell, Hawking radiation is a robust phenomenon. Whenever we do figure out quantum gravity, in other words, it should reproduce Hawking's result.

That raises a problem, one that has become notorious within theoretical physics as the *black hole information puzzle*. Remember that quantum mechanics, in its Many-Worlds version, is a deterministic theory. Randomness is only apparent, arising from self-locating uncertainty when the wave function branches and we don't know which branch we're on. But in Hawking's calculation, black-hole radiation seems not to be deterministic; it's truly random, even without any branching. Starting from a precise quantum state describing matter that collapses to make a black hole, there is no way of computing the precise quantum state of the radiation into which it evaporates. The information specifying the original state seems to be lost.

Imagine taking a book—maybe the very one you are reading right now—and throwing it into a fire, letting it burn completely away. (Don't worry, you can always buy more copies.) It might appear that the information contained in the book is lost in the flames. But if we turn on our physicist's powers of thought-experiment ingenuity, we realize that this loss is only apparent. In principle, if we captured every bit of light and heat and dust and ash from the fire, and had perfect knowledge of the laws of physics, we could reconstruct exactly what went into the fire, including all the words on the pages of the book. It'll never happen in the real world, but physics says it's conceivable.

Most physicists think that black holes should be just like that: throw a book in, and the information contained in its pages should be secretly encoded in the radiation that the black hole emits. But this is not what happens, according to Hawking's derivation of black-hole radiation; rather, the information in the book appears to be truly destroyed.

It's possible, of course, that this implication is correct, that the information really is destroyed, and that black-hole evaporation is nothing like an ordinary fire. It's not like we have any experimental input one way or the other. But most physicists believe that information is conserved, and that it really does get out somehow. And they suspect that the secret to getting it out lies in a better understanding of quantum gravity.

That's easier said than done. One way of thinking about why black holes are supposed to be black in the first place is that in order to escape, you would have to be able to travel faster than light. Hawking radiation avoids that difficulty because it actually originates right outside the event horizon, not deep in the interior. But any book we throw inside does indeed plunge into the interior, with all its information intact. You might wonder whether the information is somehow copied onto the outgoing radiation as the book falls through the horizon, and carried out that way. Unfortunately that's in contradiction with the basic principles of quantum mechanics; there is a result called the *no-cloning theorem* that says we can't duplicate quantum information without destroying the original copy.

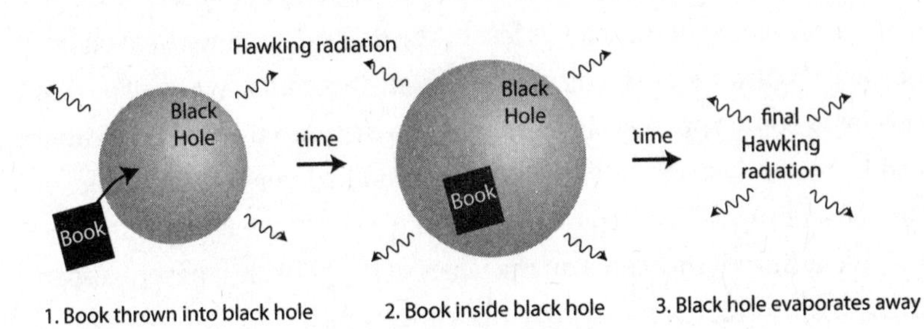

The other possibility seems to be that the book falls all the way in, but as it hits the singularity inside the black hole, its information is somehow transferred to the outgoing radiation at the horizon. Unfortunately, that would seemingly require faster-than-light communication. Or, equivalently, dynamical nonlocality—occurrences at one point in spacetime immediately influencing what happens some distance away. This kind of nonlocality is precisely what cannot happen, according to the ordinary rules of quantum field theory. This is a clue that those rules might have to be dramatically revised once quantum gravity becomes important.*

---

* It's not completely agreed upon that infalling objects actually do travel deep into the interior of a black hole. In 2012 a group of physicists argued that, if information is going to escape from evaporating black holes without violating the basic tenets of quantum mechanics, something

○ ○ ○

Hawking's proposal that black holes radiate didn't come out of the blue. It came in response to a suggestion from Jacob Bekenstein—who at the time was yet another graduate student of John Wheeler's at Princeton—that black holes should have entropy.

One of the motivations behind Bekenstein's idea was the fact that, according to classical general relativity, the area of a black hole's event horizon can never decrease. That sounds suspiciously like the second law of thermodynamics, according to which the entropy of a closed system can never decrease. Inspired by this similarity, physicists constructed an elaborate analogy between the laws of thermodynamics and the behavior of black holes, according to which the mass of the black hole is like the energy of a thermodynamic system, and the area of the event horizon is like the entropy.

Bekenstein suggested that it was more than an analogy. The area of the event horizon isn't just *like* the entropy, it *is* the entropy of the black hole, or at least proportional to it. Hawking and others scoffed at the suggestion at first—if black holes have entropy like conventional thermodynamic systems, they should also have a temperature, and then they should give off radiation! Motivated to disprove this ridiculous-sounding notion, Hawking ended up showing that it was all true. These days we refer to the entropy of a black hole as the *Bekenstein-Hawking entropy*.

One reason why this is such a provocative result is that classically, black holes don't seem like things that should have entropy at all. They're just regions of empty space. You get entropy when your system is made of atoms or other tiny constituents, which can be arranged in

---

dramatic has to happen at the event horizon: not quiet, empty spacetime, as is usually assumed, but a blast of high-energy particles known as a firewall. Opinions about the firewall proposal are divided, as theorists continue to argue back and forth about the issue.

many different ways while maintaining the same macroscopic appearance. What are these constituents supposed to be for a black hole? The answer has to come from quantum mechanics.

It's natural to presume that the Bekenstein-Hawking entropy of a black hole is a kind of entanglement entropy. There are some degrees of freedom inside the black hole, and they are entangled with the outside world. What are they?

We might first guess that the degrees of freedom are simply vibrational modes of the quantum fields inside the black hole. There are a couple of problems with that. For one thing, the real answer for the entropy of a region in quantum field theory was "infinity." We could wrestle that down to a finite number by choosing to ignore very-small-wavelength modes, but that involved introducing an arbitrary cutoff on the energies of the field vibrations we were considering. The Bekenstein-Hawking entropy, on the other hand, is just a finite number, full stop. For another thing, the entanglement entropy in field theory should depend on exactly how many fields are involved—the electrons, quarks, neutrinos, and so forth. The formula for black-hole entropy that Hawking derived makes no mention of such things at all.

If we can't simply attribute black-hole entropy to the quantum fields inside, the alternative is to imagine that spacetime itself is made of some quantum degrees of freedom, and the Bekenstein-Hawking formula measures the entanglement of the degrees of freedom inside the black hole with the degrees of freedom outside. If that sounds pretty vague, that's because it is. We're not precisely sure what these spacetime degrees of freedom are, or how they interact with one another. But the general principles of quantum mechanics should still be respected. If there's entropy, and that entropy comes from entanglement, there must be degrees of freedom that can entangle with the rest of the world in many different ways, even if classical black holes are all featureless.

If this story is right, the number of degrees of freedom in a black hole isn't infinite, but it is very large indeed. Our Milky Way galaxy contains a supermassive black hole at its center, associated with a radio source called Sagittarius A*. From observing how stars orbit around the hole, we can measure its mass to be 4 million times the mass of the sun. That corresponds to an entropy of 10^{90}, which is greater than the entropy of all the known particles in the entire observable universe. The number of degrees of freedom in a quantum system has to be at least as large as its entropy, since that entropy comes precisely from those degrees of freedom being entangled with the outside world. So there must be at least 10^{90} degrees of freedom in the black hole.

While we tend to pay attention to the stuff we see in the universe—matter, radiation, and so on—almost all of the universe's quantum degrees of freedom are invisible, doing nothing more than stitching spacetime together. In a volume of space roughly the size of an adult human, there must be at least 10^{70} degrees of freedom; we know that because that's the entropy of a black hole that would fill such a volume. But there are only about 10^{28} particles in a person. We can think of a particle as a degree of freedom that has been "turned on," while all the other degrees of freedom are peacefully "turned off" in the vacuum state. As far as quantum field theory is concerned, a human being or the center of a star isn't all that different from empty space.

◦ ◦ ◦

Maybe the fact that the entropy of a black hole is proportional to its area is just what we should expect. In quantum field theory it's natural for regions of space to have an entropy proportional to their boundary area, and a black hole is just a region of space. But a problem lurks beneath the surface. It's natural for a region of space *in the vacuum state* to have an entropy proportional to its boundary area. But a black hole isn't part of the vacuum state; there's a black hole there, and spacetime is noticeably curved.

Black holes have a very special property: they represent the *highest-*

entropy states we can have in any given size region of space. This provocative fact was first noticed by Bekenstein, and later refined by Raphael Bousso. If you start from a region within the vacuum state and try to increase its entropy, you must also increase its energy. (Since you started in the vacuum, there's nowhere for the energy to go but up.) As you keep throwing in entropy, the energy also increases. Eventually you have so much energy in a fixed region that the whole thing can't help but collapse into a black hole. That's the limit; you can't fit any more entropy into a region than you would have if a black hole were there.

That conclusion is profoundly different from what we would expect in an ordinary quantum field theory without gravity. There, there is no limit on how much entropy we can fit in a region, because there's also no limit on how much energy there can be. This reflects the fact that there are an infinite number of degrees of freedom in quantum field theory, even in a finite-sized region.

Gravity appears to be different. There is a maximum amount of energy and entropy that can fit into a given region, which seems to imply that there are only a finite number of degrees of freedom there. Somehow these degrees of freedom become entangled in the right way to stitch together into the geometry of spacetime. It's not just black holes: every region of spacetime has a maximum entropy we could imagine fitting into it (the entropy that a black hole of that size would have), and therefore a finite number of degrees of freedom. It's even true for the universe as a whole; because there is vacuum energy, the expansion of space is accelerating and that means there is a horizon all around us that delineates the extent of the observable part of our cosmos. That observable patch of space has a finite maximum entropy, so there are only a finite number of degrees of freedom needed to describe everything we see or ever will see.

If this story is on the right track, it has an immediate, profound consequence for the Many-Worlds picture of quantum mechanics. A finite number of quantum degrees of freedom implies a finite-dimensional

Hilbert space for the system as a whole (in this case, any chosen region of space). That in turn implies that there is some finite number of branches of the wave function, not an infinite number. That's why Alice was cagey back in Chapter Eight about whether there are an infinite number of "worlds" in the wave function. In many simple models of quantum mechanics, including that of a fixed set of particles moving smoothly through space or any ordinary quantum field theory, Hilbert space is infinite-dimensional and there could potentially be an infinite number of worlds. But gravity seems to change things around in an important way. It prevents most of those worlds from existing, because they would describe too much energy being packed into a local region.

So maybe in the real universe, where gravity certainly exists, Everettian quantum mechanics only describes a finite number of worlds. The number Alice mentioned for the dimensionality of Hilbert space was $2^{10^{122}}$.

Now we can reveal where that number came from: it's from calculating the entropy that our observable universe will have once it reaches maximum entropy, and working backward to find out how big Hilbert space needs to be to accommodate that much entropy. (The size of the observable universe is set by the vacuum energy, so the exponent 10^{122} is the ratio of the Planck scale to the cosmological constant, familiar from our discussion in Chapter Twelve.) Our confidence in the basic principles of quantum gravity isn't strong enough to be absolutely sure that there are only a finite number of Everettian worlds, but it seems reasonable, and it certainly would make things much simpler.

○ ○ ○

The maximum-entropy nature of black holes also has an important consequence for quantum gravity. In classical general relativity, there's nothing special about the interior region of a black hole, in between the event horizon and the singularity. There's a gravitational field there, but to an infalling observer it otherwise looks like empty space. According to the story we told in the last chapter, the quantum version of "empty space" is something like "a collection of spacetime degrees of freedom entangled together in such a way as to form an emergent three-dimensional geometry." Implicit in that description is that the degrees of freedom are scattered more or less uniformly throughout the volume of space we're looking at. And if that were true, the maximum-entropy state of that form would have all of those degrees of freedom entangled with the outside world. The entropy would thus be proportional to the volume of the region, not the area of its boundary. What's up?

There is a clue from the black hole information puzzle. The issue there was that there is no obvious way to transmit information from a book that has fallen into the black hole to the Hawking radiation emitted from the event horizon, at least not without signals moving faster than light. So what about this crazy idea: maybe all of the information about the state of the black hole—the "inside" as well as the horizon—can be thought of as living on the horizon itself, not buried in the interior. The black-hole state "lives," in some sense, on a two-dimensional surface, rather than being stretched across a three-dimensional volume.

First developed by Gerard 't Hooft and Leonard Susskind in the 1990s, based in part on a paper by Charles Thorn from 1978, this idea is known as the *holographic principle*. In an ordinary hologram, shining light on a two-dimensional surface reveals an apparently three-dimensional image. According to the holographic principle, the apparently three-dimensional interior of a black hole reflects information encoded on the two-dimensional surface of its event horizon. If this is true, maybe it's not so hard to get information from the black hole to its outgoing radiation, because the information was always on the horizon to start with.

Physicists still haven't settled on the precise meaning of holography for real-world black holes. Is it just a way of counting the number of

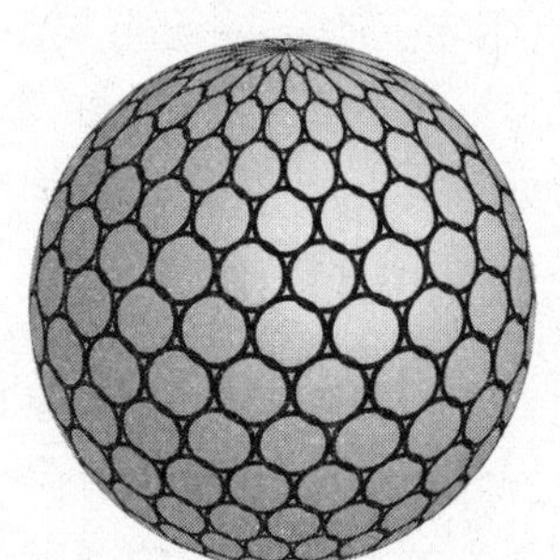

degrees of freedom, or should we think that there is an actual two-dimensional theory living on the event horizon that describes the physics of the black hole? We don't know, but there is a different context in which holography is very precise: the so-called *AdS/CFT correspondence*, proposed by Juan Maldacena in 1997. The "AdS" in the label stands for "anti–de Sitter space," a hypothetical spacetime with no matter sources other than a negative vacuum energy (as opposed to the positive vacuum energy of our real world). "CFT" stands for *conformal field theory*, a particular kind of quantum field theory that can be defined on an infinitely faraway boundary of AdS. According to Maldacena, these two theories are secretly equivalent to each other. That's extremely provocative, for a couple of reasons. First, the AdS theory includes gravity, while the CFT is an ordinary field theory that has no gravity at all. Second, the boundary of a spacetime has one fewer dimensions than the spacetime itself. If we consider four-dimensional AdS, for example, that is equivalent to a three-dimensional conformal field theory. You couldn't ask for a more explicit example of holography in action.

Going into the details of AdS/CFT would require another book entirely. But it is worth mentioning that it is in this context that most modern research on the connection between spacetime geometry and quantum entanglement is being carried out. As noted by Shinsei Ryu, Tadashi Takayanagi, Mark Van Raamsdonk, Brian Swingle, and others in the early 2000s, there is a direct connection between entanglement in the boundary CFT and the resulting geometry in the AdS interior. Because AdS/CFT is relatively well defined as models of quantum gravity go, understanding this connection has been the target of a very intense effort over the past several years.

Alas, it's not the real world. All of the fun of AdS/CFT comes from relating things in the interior, where gravity happens, to things on the boundary, where gravity is absent. But the existence of the boundary is very special to anti–de Sitter space, which relies on a negative vacuum energy. Our universe appears to have a positive vacuum energy, not a negative one.

There's an old joke about the drunk who is looking under a lamppost for his lost keys. When someone asks if he's sure he lost them there, he replies, "Oh no, I lost them somewhere else, but the light is much better over here." In the quantum-gravity game, AdS/CFT is the world's brightest lamppost. By studying it we've uncovered a large number of fascinating concepts that are useful to theoretical physicists, but there is no direct route to using that knowledge to understand why apples fall from trees, or other aspects of gravity in the space around us. It's worth continuing the pursuit, but important to keep our eyes on the prize: understanding the world in which we actually live.

∘ ∘ ∘

The implications of holography for real-world black holes are less clear than they are for the imaginary world of AdS/CFT. Are we saying that classical general relativity was completely wrong about the interior of a black hole appearing empty, and that in fact an infalling observer would smack into a holographic surface upon encountering the event horizon? We are not—at least, most adherents of holography aren't saying that. Rather, they appeal to a related and equally startling idea, *black-hole complementarity*. It was proposed by Susskind and others, using terminology that intentionally recalls Bohr's philosophy of quantum measurement.

The black-hole version of complementarity says that things are a little more nuanced than simply "the interior of a black hole looks like ordinary empty space" or "all the information about the black hole is encoded on the event horizon." In fact both are true, but we can't speak both languages at the same time. Or, as physicists are more likely to put it, they don't simultaneously appear true to any single observer. To an observer falling through the event horizon, everything looks like normal empty space, while to an observer looking at the hole from far away, all of the information is spread across the horizon.

Even though this behavior is fundamentally quantum-mechanical, it does have a classical precursor. Think about what happens to a book (or a star, or whatever) when we throw it into a black hole in classical general relativity. From the book's point of view, it just passes right into the interior. But the effect of spacetime warping is strong near the event horizon, so that's not what an external observer would see. They would see the book appear to slow down as it approached the horizon, becoming redder and dimmer along the way. They wouldn't ever see it cross; to someone far away, objects appear to be frozen in time as they approach the horizon, rather than plunging in. This led astrophysicists to develop a picture called the *membrane paradigm*, according to which we can model the physical properties of a black hole by imagining that there is a physical membrane at the horizon, with certain calculable properties such as temperature and electrical conductivity. The membrane paradigm was originally thought of as a convenient shortcut through which astrophysicists could simplify calculations involving black holes, but complementarity claims that external observers really do see black holes as if they were vibrating quantum membranes where the classical event horizon would be.

If you tend to think of spacetime as a fundamental thing, this might make no sense at all. Spacetime has some geometry, there's nothing else to it. But quantum-mechanically it's perfectly plausible; there's a wave function of the universe, and different observations can reveal different things about it. It's not that much different from saying that the number of particles in a state depends on how we observe it.

The world is a quantum state evolving in Hilbert space, and physical space emerges out of that. It shouldn't come as a surprise that a single quantum state might exhibit different notions of position and locality depending on what kind of observations we perform on it. According to black-hole complementarity, there's no such thing as "what the geometry of spacetime is," or "where the degrees of freedom are"; you ask either what the quantum state is, or what is seen by some particular observer.

This sounds different from the picture we explored in the last chapter, where degrees of freedom were distributed in a network filling space, and became entangled to define an emergent geometry. But that picture was only meant to apply when gravity was weak, and black holes definitely do not qualify as weak. In the view presented in this chapter, there are still abstract degrees of freedom coming together to form spacetime, but "where they are located" depends on how they are being observed. Space itself is not fundamental; it's just a useful way of talking from certain points of view.

∘ ∘ ∘

Hopefully these last chapters have successfully conveyed the way in which Many-Worlds quantum mechanics might have significant implications for the long-standing problem of quantum gravity. To be honest, many physicists working on these problems don't think of themselves as using Many-Worlds, though they are implicitly doing so. They certainly are not using hidden variables, or dynamical collapses, or an epistemic approach to quantum mechanics. When it comes to understanding how to quantize the universe itself, Many-Worlds seems to be the most direct path to take, if nothing else.

Is the picture we've sketched, where the entanglement between degrees of freedom somehow comes together to define the geometry of our approximately classical spacetime, actually on the right track? Nobody knows for sure. What seems clear, given the current state of our knowledge, is that both space and time could emerge from an abstract quantum state in the desired way—all the ingredients are there, and it's not out of place to hope that a few more years of work will bring a much sharper picture into focus. If we train ourselves to discard our classical prejudices, and take the lessons of quantum mechanics at face value, we may eventually learn how to extract our universe from the wave function.

EPILOGUE

Everything Is Quantum

What would Einstein have thought of Many-Worlds quantum theory? Likely he would have been repulsed, at least at first exposure. But he would have to admit that there are aspects of the idea that fit very well with his picture of how nature should operate.

Einstein died in Princeton in 1955, just as Everett was wrangling his idea into shape. He was firmly committed to the principle of locality, and was enormously bothered by the spooky action at a distance implied by quantum entanglement. In that sense, he might very well have been horrified by Many-Worlds and the holographic principle, ideas that treat space itself as emergent rather than fundamental. The suggestion that reality is described as a vector in an enormous Hilbert space, rather than as matter and energy in good old four-dimensional spacetime, is not one he would have found congenial. But there's a good chance that he would have been pleased that Everett returns our best description of the universe to one featuring definite, deterministic evolution—and reaffirms the principle that reality is ultimately knowable.

Late in life, Einstein related a story from his childhood.

> A wonder of this kind I experienced as a child of four or five years when my father showed me a compass. That this needle behaved in such a determined way did not at all fit in the kind of occurrences that could find a place in the unconscious world of concepts (efficacy produced by direct "touch"). I can still remember—or at least believe I can remember—that this experience made a deep and lasting impression upon me.
>
> Something deeply hidden had to be behind things.

It seems to me that this impulse lies at the heart of all of Einstein's worries about quantum mechanics. He might have fretted out loud about indeterminism and nonlocality, but what really bugged him was his sense that Copenhagen quantum mechanics replaced the crisp rigor of good scientific theories with a fuzzy paradigm in which an ill-defined notion of "measurement" played a central role. He was always on the lookout for the deeply hidden thing beneath the surface, the principle that would restore intelligibility to that which had drifted into mystery. Little did he suspect that what was hidden might be other branches of the wave function.

It doesn't really matter what Einstein would have actually thought, of course; scientific theories rise or fall on their merits, not because we can conjure up hypothetical ghosts of great minds from the past to nod their approval.

But it's useful to pay attention to those great minds, if only to be reminded of the connections between debates of the past and research in the present. The issues discussed in this book stem directly from the discussions between Einstein and Bohr and others in the 1920s. In the wake of the Solvay Conference, popular opinion within the physics community swung Bohr's way, and the Copenhagen approach to quantum mechanics settled in as entrenched dogma. It's proven to be an amazingly successful tool at making predictions for experiments and designing new technologies. But as a fundamental theory of the world, it falls woefully short.

I've laid out the case for why Many-Worlds is the most promising formulation of quantum mechanics. But I have enormous respect for, and have frequent productive conversations with, partisans for other approaches. What makes me melancholy are professional physicists who dismiss foundational work and don't think the issues are worth taking seriously. After reading this book, whether or not you would describe yourself as an Everettian, I hope you are convinced of the importance of getting quantum mechanics right once and for all.

I'm optimistic about how things are progressing. The modern study of quantum foundations isn't just a bunch of elderly physicists chatting about fantastical ideas over tumblers of scotch after the real work is done for the day. Much of the recent progress in developing our understanding of quantum theory has been spurred, directly or indirectly, by technological innovations: quantum computing, quantum cryptography, and quantum information more generally. We've reached a point where it is no longer practical to draw a bright line between the quantum and classical realms. Everything is quantum. This state of affairs has forced physicists to take the foundations of quantum mechanics a bit more seriously, and has led to new insights that might help explain the emergence of space and time themselves.

I think we'll be making significant progress on these difficult puzzles in the near future. And I like to believe most of the other versions of me on other branches of the wave function feel likewise.

APPENDIX

The Story of Virtual Particles

Our discussion of quantum field theory in Chapter Twelve would seem amusingly idiosyncratic to most working quantum field theorists. What we cared about was just the vacuum state, the lowest-energy configuration of a set of quantum fields filling space. But that's just one state out of an infinite number. What most physicists care about are all the other states—those that look like particles moving and interacting with one another.

Just as it's natural to speak about "the position of the electron" when we really know better and should speak about the electron's wave function, physicists who understand perfectly well that the world is made of fields tend to talk about particles all the time. They even call themselves "particle physicists" without discernible embarrassment. It's an understandable impulse: particles are what we see, regardless of what's going on beneath the surface.

The good news is, that's okay, as long as we know what we're doing. For many purposes, we can talk *as if* what really exists is a collection of particles traveling through space, bumping into one another, being created and destroyed, and occasionally popping into or out of existence.

The behavior of quantum fields can, under the right circumstances, be accurately modeled as the repeated interaction of many particles. That might seem natural when the quantum state describes some fixed number of particle-like field vibrations, far away from one another and blissfully unaware of the others' existence. But if we follow the rules, we can calculate what happens using particle language even when a bunch of fields are vibrating right on top of one another, exactly when you might expect their field-ness to be most important.

That's the essential insight from Richard Feynman and his well-known tool of *Feynman diagrams*. When he first invented his diagrams, Feynman held out the hope that he was suggesting a particle-based alternative to quantum field theory, but that turns out not to be the case. What they are is both a wonderfully vivid metaphorical device and an incredibly convenient computational method, within the overarching paradigm of quantum field theory.

A Feynman diagram is simply a stick-figure cartoon representing particles moving and interacting with one another. With time running from left to right, an initial set of particles comes in, they jumble up with various particles appearing or disappearing, then a final set of particles emerges. Physicists use these diagrams not only to describe what processes are allowed to happen but to precisely calculate the likelihood that they actually will. If you want to ask, for example, what particles a Higgs boson might decay into and how rapidly, you would do a calculation involving a boatload of Feynman diagrams, each representing a certain contribution to the final answer. Likewise if you want to know how likely it is that an electron and a positron will scatter off each other.

Here is a simple Feynman diagram. The way to think about this picture is that an electron and a positron (straight lines) come in from the left, meet each other, and annihilate into a photon (wavy line), which travels for a while before converting back into an electron/positron pair. There are specific rules that allow physicists to attach precise numbers

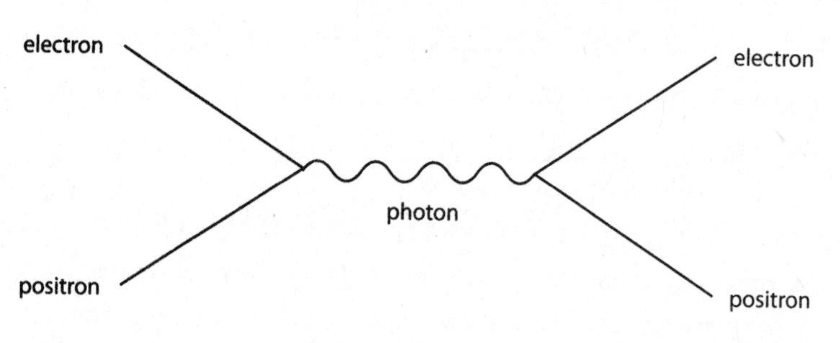

to every such diagram, indicating the contribution that this picture makes to the overall process of "an electron and a positron scatter off each other."

The story we tell based on the Feynman diagrams is just that, a story. It's not literally true that an electron and a positron change into a photon and then change back. For one thing, real photons move at the speed of light, while electron/positron pairs (either the individual particles or the center of mass of a pair of them) do not.

What actually happens is that both the electron field and the positron field are constantly interacting with the electromagnetic field; oscillations in any electrically charged field, such as the electron or positron, are necessarily accompanied by subtle oscillations in the electromagnetic field as well. When the oscillations in two such fields (which we interpret as the electron and positron) come close to each other or overlap, all of the fields push and pull on one another, causing our original particles to scatter off in some direction. Feynman's insight is that we can calculate what's going on in the field theory by pretending that there are a bunch of particles flying around in certain ways.

This represents an enormous computational convenience; working particle physicists use Feynman diagrams all the time, and occasionally dream about them while sleeping. But there are certain conceptual compromises that need to be made along the way. The particles confined to the interior of the Feynman diagrams, which don't either come in from the left or exit to the right, don't obey the usual rules for ordinary particles. They don't, for example, have the same energy or mass that a regular particle has. They obey their own set of rules, just not the usual ones.

That shouldn't be surprising, as the "particles" inside Feynman diagrams are not particles at all; they're a convenient mathematical fairy tale. To remind ourselves of that, we label them "virtual" particles. Virtual particles are just a way to calculate the behavior of quantum fields, by pretending that ordinary particles are changing into weird particles with impossible energies, and tossing such particles back and forth between themselves. A real photon has exactly zero mass, but the mass of a virtual photon can be absolutely anything. What we mean by "virtual particles" are subtle distortions in the wave function of a collection of quantum fields. Sometimes they are called "fluctuations" or simply "modes" (referring to a vibration in a field with a particular wavelength). But everyone calls them particles, and they can be successfully represented as lines within Feynman diagrams, so we can call them that.

◦ ◦ ◦

The diagram we drew for an electron and a positron scattering off each other isn't the only one we could possibly draw; in fact, it's just one of an infinite number. The rules of the game tell us that we should sum up all of the possible diagrams with the same incoming and outgoing particles. We can list such diagrams in order of increasing complexity, with subsequent diagrams containing more and more virtual particles.

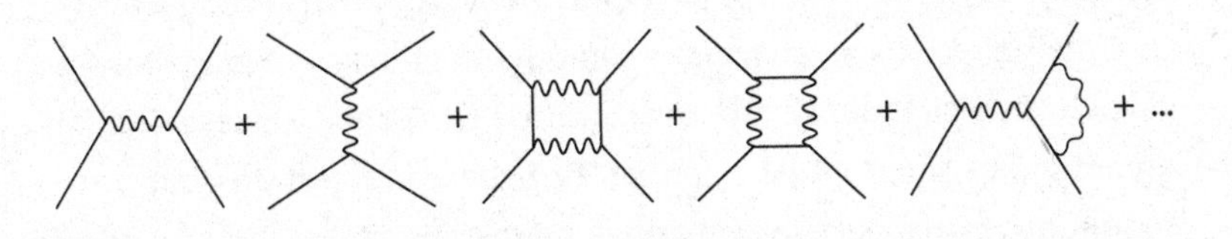

The final number we obtain is an amplitude, so we square it to get the probability of such a process happening. Using Feynman diagrams, we can calculate the probability of two particles scattering off each other, of one particle decaying into several, or for particles turning into other kinds of particles.

An obvious worry pops up: If there are an infinite number of diagrams, how can you add them all up and get a sensible result? The answer is that diagrams contribute smaller and smaller amounts as they become more complicated. Even though there are an infinite number of them, the sum total of all the very complicated ones can be a tiny number. In practice, as a matter of fact, we often get quite accurate answers by calculating only the first few diagrams in the infinite series.

There is one subtlety along the way to this nice result, however. Consider a diagram that has a loop in it—that is, where we can trace around some set of particle lines to form a closed circle. Here is an electron and a positron exchanging two photons:

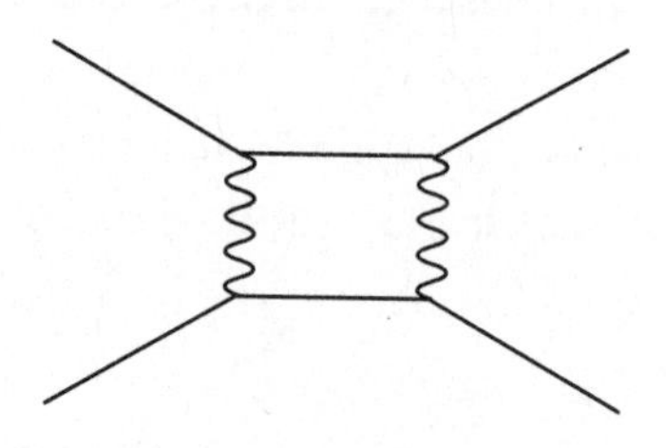

Each line represents a particle with a certain amount of energy. This energy is conserved when lines come together: if one particle comes in and splits into two, for example, the sum of the energies of those two particles must equal that of the initial particle. But how that energy gets split up is completely arbitrary, as long as the sum total is fixed. In fact, due to the wacky logic of virtual particles, the energy of one particle can even be a negative number, such that the other one has more energy than the initial particle did.

This means that when we calculate the process described by a Feynman diagram with an internal closed loop, an arbitrarily large amount of energy can be traveling down any particular line within the loop.

Sadly, when we do the calculation for what such diagrams contribute to the final answer, the result can turn out to be infinitely large. That's the origin of the infamous infinities plaguing quantum field theory. Obviously the probability of a certain interaction can be at most 1, so an infinite answer means we've taken a wrong turn somehow.

Feynman and others managed to work out a procedure for dealing with these infinities, now known as *renormalization*. When you have a bunch of quantum fields that interact with one another, you can't simply first treat them separately, and then add in the interactions at the end. The fields are constantly, inevitably affecting one another. Even when we have a small vibration in the electron field, which we might be tempted to identify as a single electron, there are inevitably accompanying vibrations in the electromagnetic field, and indeed in all the other fields that the electron interacts with. It's like playing a piano note in a showroom with many pianos present; the other instruments will begin to gently hum along with the original one, causing a faint echo of whatever notes you are playing. In Feynman-diagram language, this means that even an isolated particle propagating through space is actually accompanied by a surrounding cloud of virtual particles.

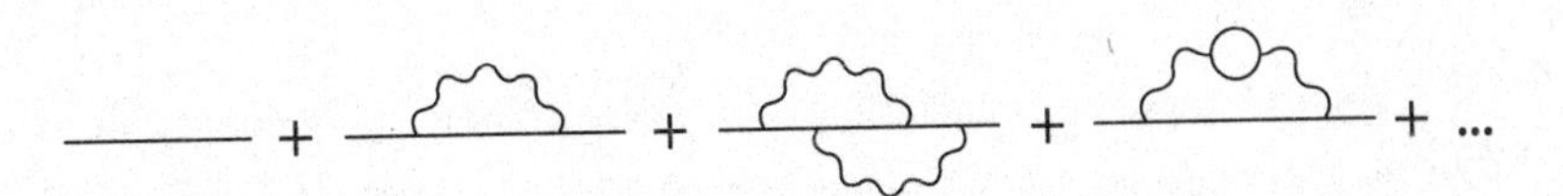

As a result, it's helpful to distinguish between the "bare" fields as they would behave in an imaginary world where all interactions were simply turned off, and the "physical" fields that are accompanied by other fields they interact with. The infinities that you get by naïvely turning a crank in the Feynman diagrams are simply a result of trying to work with bare fields, whereas what we really observe are physical ones. The adjustment required to go from one to another is sometimes informally described as "subtracting off infinity to get a finite answer,"

but that's misleading. No physical quantities are infinite, nor were they ever; the infinities that quantum-field-theory pioneers managed to "hide" were simply artifacts of the very big difference between fields that interact and fields that don't. (We face exactly this kind of issue when trying to estimate the vacuum energy in quantum field theory.)

Nevertheless, renormalization comes with important physical insights. When we want to measure some property of a particle, such as its mass or charge, we probe it by seeing how it interacts with other particles. Quantum field theory teaches us that the particles we see aren't simple point-like objects; each particle is surrounded by a cloud of other virtual particles, or (more accurately) by the other quantum fields it interacts with. And interacting with a cloud is different from interacting with a point. Two particles that smash into each other at high velocity will penetrate deep into each other's clouds, seeing relatively compact vibrations, while two particles that pass by slowly will see each other as (relatively) big puffy balls. Consequently, the apparent mass or charge of a particle will depend on the energy of the probes with which we look at it. This isn't just a song and dance: it's an experimental prediction, which has been seen unmistakably in particle-physics data.

○ ○ ○

The best way to think about renormalization wasn't really appreciated until the work of Nobel laureate Kenneth Wilson in the early 1970s. Wilson realized that all of the infinities in Feynman-diagram calculations came from virtual particles with very large energies, corresponding to processes at extremely short distances. But high energies and short distances are precisely where we should have the least confidence that we know what's going on. Processes with very high energies could involve completely new fields, ones that have such high masses that we haven't yet produced them in experiments. For that matter, spacetime itself might break down at short distances, perhaps at the Planck length.

So, Wilson reasoned, what if we're just a little bit more honest, and admit that we don't know what's going on at arbitrarily high energies? Instead of taking loops in Feynman diagrams and allowing the energies of the virtual particles to go up to infinity, let's include an explicit cutoff in the theory: an energy above which we don't pretend to know what's happening. The cutoff is in some sense arbitrary, but it makes sense to put it at the dividing line between energies about which we have good experimental knowledge, and above which we haven't been able to peek. There can even be a physically good reason to choose a certain cutoff, if we expect new particles or other phenomena to kick in at that scale, but don't know exactly what they will be.

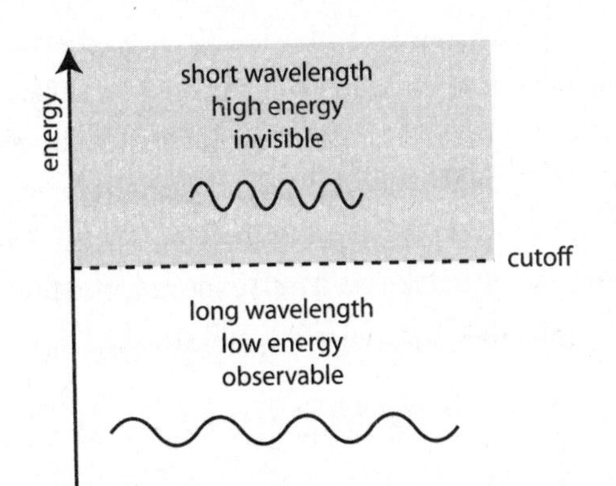

Of course, there could be interesting things going on at higher energies, so by including a cutoff we're admitting that we're not getting exactly the right answer. But Wilson showed that what we do get is generally more than good enough. We can precisely characterize how, and roughly by how much, any new high-energy phenomena could possibly affect the low-energy world we actually see. By admitting our ignorance in this way, what we're left with is an *effective field theory*—one that doesn't presume to be an exact description of anything, but one that can successfully fit the data we actually have. Modern quantum field theorists recognize that all of their best models are actually effective field theories.

This leaves us with a good news/bad news situation. The good news is that we are able to say an enormous amount about the behavior of particles at low energies, using the magic of effective field theory, even if we don't know everything (or anything) about what's happening at higher energies. We don't need to know all the final answers in order to say something reliable and true. That's a big part of why we can be confident that the laws of physics governing the particles and forces that make up you and me and our everyday environments are completely known: those laws take the form of an effective field theory. There's plenty of room to discover new particles and forces, but either they must be too massive (high energy) to have yet been produced in experiments, or they interact with us so incredibly weakly that they can't possibly have an effect on tables and chairs and cats and dogs and other pieces of the architecture of our low-energy world.

The bad news is that we would very much like to learn more about what's really going on at high energies and short distances, but the magic of effective field theory makes that extremely hard. It's good that we can accurately describe low-energy physics no matter what is going on at higher energies, but it's also frustrating because this seems to imply that we can't infer what's going on up there without somehow probing it directly. This is why particle physicists are so enamored of building ever larger and higher-energy particle accelerators; that's the only reliable way we know of to discover how the universe works at very small distances.

ACKNOWLEDGMENTS

Every book is a collaboration, and this one more than others. There is much to be said about quantum mechanics, and there was definitely a temptation to say it all. That might have been a fun book to write but it would have been a tedious chore to read. I owe a variety of generous and insightful readers for their help in wrestling the manuscript down to something manageable and, hopefully, in parts, fun. I should specifically mention helpful comments from Nick Aceves, Dean Buonomano, Joseph Clark, Don Howard, Jens Jäger, Gia Mora, Jason Pollack, Daniel Ranard, Rob Reid, Grant Remmen, Alex Rosenberg, Landon Ross, Chip Sebens, Matt Strassler, and David Wallace. In ways stretching from small—offhandedly mentioning something in conversation that later ended up in the book—to large—reading every chapter and offering useful insights—these generous folks helped rescue me from writing a book that would not have been nearly as good.

I want to give special thanks to Scott Aaronson, who is the best test-reader a physicist/author could ask for, giving the text a thorough reading and offering invariably useful feedback on both substance and style. I'll also mention Gia Mora again, because she was inexplicably omitted from the acknowledgments of *The Big Picture*, and I feel bad about that.

It goes without saying that I've learned an enormous amount about quantum mechanics and spacetime from a large number of extremely smart people over the years, and their influence pervades this book even if I didn't talk specifically about the words written here. Many thanks go to David Albert, Ning Bao, Jeff Barrett, Charles Bennett, Adam Becker, Kim Boddy, Charles Cao, Aidan Chatwin-Davies, Sidney Coleman, Edward Farhi, Alan Guth, James Hartle, Jenann Ismael, Matthew Leifer, Seth Lloyd, Frank Maloney, Tim Maudlin, Spiros Michalakis, Alyssa Ney, Don Page, Alain Phares, John Preskill, Jess Reidel, Ashmeet Singh, Leonard Susskind, Lev Vaidman, Robert Wald, and Nicholas Warner, not to mention the numerous others I am doubtless forgetting.

Thanks as usual to my students and collaborators for tolerating my occasional absences while trying to finish the book. And thanks also to the students in 125C, the third quarter of Caltech's course on quantum mechanics for juniors, who tolerated me teaching them about decoherence and entanglement rather than just the familiar routine of solving the Schrödinger equation over and over.

A million thanks to my editor at Dutton, Stephen Morrow, whose patience and insight were more sorely needed for this book than they have been in the past. He even let me include an entire chapter in dialogue form, although it's possible I just wore him down. An author couldn't imagine an editor who cared more about the quality of the final product, and much of the quality here is due to Stephen. Thanks also to my agents, Katinka Matson and John Brockman, who always make a process that could potentially be nerve-racking into something tolerable, possibly even enjoyable.

And the most thanks of all to Jennifer Ouellette, the perfect partner in writing and in life. Not only did she support me in countless ways along the journey, but she took time out of her own very demanding writing schedule to go through every page here carefully and offer invaluable insight and tough love. I didn't delete nearly as much as she suggested, and probably the book is poorer thereby, but trust me, it's way better than it was before she got to it.

Thanks also to Jennifer for bringing into our lives Ariel and Caliban, the best writing-partner cats an author could ask for. No actual cats were subjected to thought experiments during the composition of this book.

FURTHER READING

There have obviously been a large number of books written about quantum mechanics. Here are a few that are relevant to the themes of this book:

Albert, D. Z. (1994). *Quantum Mechanics and Experience.* Harvard University Press. A short introduction to quantum mechanics and the measurement problem from a philosophical perspective.

Becker, A. (2018). *What Is Real? The Unfinished Quest for the Meaning of Quantum Physics.* Basic Books. A historical overview of quantum foundations, including alternatives to Many-Worlds and the obstacles that many physicists faced in thinking about these issues.

Deutsch, D. (1997). *The Fabric of Reality.* Penguin. An introduction to Many-Worlds but also much more, from computation to evolution to time travel.

Saunders, S., J. Barrett, A. Kent, and D. Wallace. (2010). *Many Worlds? Everett, Quantum Theory, and Reality.* A collection of essays for and against Many-Worlds.

Susskind, L., and A. Friedman. (2015). *Quantum Mechanics: The Theoretical Minimum.* Basic Books. A serious introduction to quantum mechanics, taught at the level of an introductory course for physics students at a good university.

Wallace, D. (2012). *The Emergent Multiverse: Quantum Theory According to the Everett Interpretation.* Oxford University Press. Somewhat technical, but this is the now-standard reference book on Many-Worlds.

REFERENCES

Prologue

Don't Be Afraid

"I think I can safely say": See R. P. Feynman (1965), *The Character of Physical Law*, MIT Press, 123.

Chapter 2

The Courageous Formulation

Austere Quantum Mechanics

"Shut up and calculate": See N. D. Mermin (2004), "Could Feynman Have Said This?" *Physics Today* 57, 5, 10.

Chapter 3

Why Would Anybody Think This?

How Quantum Mechanics Came to Be

"six impossible things": L. Carroll (1872), *Through the Looking Glass and What Alice Found There*, Dover, 47.

"Sweet is by convention": Quoted in H. C. Von Baeyer (2003), *Information: The New Language of Science*, Weidenfeld & Nicolson, 12.

"very revolutionary": Quoted in R. P. Crease and A. S. Goldhaber (2014), *The Quantum Moment: How Planck, Bohr, Einstein, and Heisenberg Taught Us to Love Uncertainty*, W. W. Norton & Company, 38.

"There appears to me one grave difficulty": Quoted in H. Kragh (2012), "Rutherford, Radioactivity, and the Atomic Nucleus," https://arxiv.org/abs/1202.0954.

"had written a crazy paper": Quoted in A. Pais (1991), *Niels Bohr's Times, in Physics, Philosophy, and Polity*, Clarendon Press, 278.

"A veritable sorcerer's calculation": Quoted in J. Bernstein (2011), "A Quantum Story," *The Institute Letter*, Institute for Advanced Study, Princeton.

"I don't like it": Quoted in J. Gribbin (1984), *In Search of Schrödinger's Cat: Quantum Physics and Reality*, Bantam Books, v.

Chapter 4

What Cannot Be Known, Because It Does Not Exist

Uncertainty and Complementarity

For more on the double-slit experiment, see A. Ananthaswamy (2018), *Through Two Doors at Once: The Elegant Experiment That Captures the Enigma of Our Quantum Reality*, Dutton.

Chapter 5

Entangled Up in Blue

Wave Functions of Many Parts

A. Einstein, B. Podolsky, and N. Rosen (1935), "Can Quantum-Mechanical Description of Reality Be Considered Complete?" *Physical Review* 47, 777.

For general insight into Bell's theorem and its relationship to EPR and Bohmian mechanics, see T. Maudlin (2014), "What Bell Did," *Journal of Physics* A 47, 424010.

"secular press": Quoted in W. Isaacson (2007), *Einstein: His Life and Universe*, Simon & Schuster, 450.

D. Rauch et al. (2018), "Cosmic Bell Test Using Random Measurement Settings from High-Redshift Quasars," *Physical Review Letters* 121, 080403.

Chapter 6

Splitting the Universe

Decoherence and Parallel Worlds

A good biography of Hugh Everett is P. Byrne (2010), *The Many Worlds of Hugh Everett III: Multiple Universes, Mutual Assured Destruction, and the Meltdown of a*

Nuclear Family, Oxford University Press. Quotes in this chapter are largely from this book and A. Becker (2018), *What Is Real?*, Basic Books.

Everett's original paper (long and short versions) and various commentaries can be found in B. S. DeWitt and N. Graham (1973), *The Many Worlds Interpretation of Quantum Mechanics*, Princeton University Press.

"Nothing has done more to convince me": Quoted in A. Becker (2018), *What Is Real?*, Basic Books, 127.

H. D. Zeh (1970), "On the Interpretation of Measurements in Quantum Theory," *Foundations of Physics* 1, 69.

"The Copenhagen Interpretation is hopelessly incomplete": Quoted in P. Byrne (2010), 141.

"Split?": Quoted in P. Byrne (2010), 139.

"Lest the discussion of my paper die": Quoted in P. Byrne (2010), 171.

"doomed from the beginning": Quoted in A. Becker (2018), 136.

"I can't resist asking": Quoted in P. Byrne (2010), 176.

"I realize that there is a certain value": M. O. Everett (2007), *Things the Grandchildren Should Know*, Little, Brown, 235.

Chapter 7

Order and Randomness

Where Probability Comes From

"Why do people say": Quoted in G.E.M. Anscombe (1959), *An Introduction to Wittgenstein's Tractatus*, Hutchinson University Library, 151.

"fatness measure": D. Z. Albert (2015), *After Physics*, Harvard University Press, 169.

W. H. Zurek (2005), "Probabilities from Entanglement, Born's Rule from Envariance," *Physical Review* A 71, 052105.

C. T. Sebens and S. M. Carroll (2016), "Self-Locating Uncertainty and the Origin of Probability in Everettian Quantum Mechanics," *British Journal for the Philosophy of Science* 69, 25.

D. Deutsch (1999), "Quantum Theory of Probability and Decisions," *Proceedings of the Royal Society of London* A455, 3129.

For a comprehensive review of the decision-theoretic approach to the Born rule, see D. Wallace (2012), *The Emergent Multiverse*.

Chapter 8

Does This Ontological Commitment Make Me Look Fat?

A Socratic Dialogue on Quantum Puzzles

"mistaken and even a vicious": K. Popper (1967), "Quantum Mechanics Without the Observer," in M. Bunge (ed.), *Quantum Theory and Reality. Studies in the Foundations Methodology and Philosophy of Science*, vol. 2, Springer, 12.

"a completely objective discussion": K. Popper (1982), *Quantum Theory and the Schism in Physics*, Routledge, 89.

For more on entropy and the arrow of time, see S. M. Carroll (2010), *From Eternity to Here: The Quest for the Ultimate Theory of Time*, Dutton.

"Asking how many worlds": D. Wallace (2012), *The Emergent Multiverse*, 102.

"Despite the unrivaled empirical success": D. Deutsch (1996), "Comment on Lockwood," *British Journal for the Philosophy of Science* 47, 222.

Chapter 9

Other Ways

Alternatives to Many-Worlds

"clearly being assigned": Quoted in A. Becker (2018), *What Is Real?*, Basic Books, 213.

"If we cannot disprove Bohm": Quoted in A. Becker (2018), 90.

"the paper is completely senseless": Quoted in A. Becker (2018), 199.

"Everett phone": J. Polchinski (1991), "Weinberg's Nonlinear Quantum Mechanics and the Einstein-Podolsky-Rosen Paradox," *Physical Review Letters* 66, 397.

For more on hidden-variable and dynamical-collapse models, see T. Maudlin (2019), *Philosophy of Physics: Quantum Theory*, Princeton.

R. Penrose (1989), *The Emperor's New Mind: Concerning Computers, Minds, and the Laws of Physics*, Oxford.

"the Einstein-Podolsky-Rosen paradox is resolved": J. S. Bell (1966), "On the Problem of Hidden-Variables in Quantum Mechanics," *Reviews of Modern Physics* 38, 447.

"a superfluous ideological superstructure," and **"artificial metaphysics"**: Quoted in W. Myrvold (2003), "On Some Early Objections to Bohm's Theory," *International Studies in the Philosophy of Science* 17, 7.

H. C. Von Baeyer (2016), *QBism: The Future of Quantum Physics*, Harvard.

"There is indeed" and **"QBism regards"**: N. D. Mermin (2018), "Making Better Sense of Quantum Mechanics," *Reports on Progress in Physics* 82, 012002.

C. A. Fuchs (2017), "On Participatory Realism," in I. Durham and D. Rickles, eds., *Information and Interaction*, Springer.

"The Everett interpretation (insofar as it is philosophically acceptable)": D. Wallace (2018), "On the Plurality of Quantum Theories: Quantum Theory as a Framework, and Its Implications for the Quantum Measurement Problem," in S. French and J. Saatsi, eds., *Scientific Realism and the Quantum*, Oxford.

Chapter 10

The Human Side

Living and Thinking in a Quantum Universe

M. Tegmark (1998), "The Interpretation of Quantum Mechanics: Many Worlds or Many Words?" *Fortschrift Physik* 46, 855.

R. Nozick (1974), *Anarchy, State, and Utopia*, Basic Books, 41.

"All that quantum mechanics purports to provide": E. P. Wigner (1961), "Remarks on the Mind-Body Problem," in I. J. Good, *The Scientist Speculates*, Heinemann.

Chapter 11

Why Is There Space?

Emergence and Locality

I talk more about emergence (and the Core Theory) in S. M. Carroll (2016), *The Big Picture: On the Origins of Life, Meaning, and the Universe Itself*, Dutton.

"I think my father": James Hartle (2016), personal communication.

Chapter 12

A World of Vibrations

Quantum Field Theory

"It is inconceivable that inanimate brute matter should": I. Newton (2004), *Newton: Philosophical Writings*, ed. A. Janiak, Cambridge, 136.

P.C.W. Davies (1984), "Particles Do Not Exist," in B. S. DeWitt, ed., *Quantum Theory of Gravity: Essays in Honor of the 60th Birthday of Bryce DeWitt*, Adam Hilger.

Chapter 13

Breathing in Empty Space

Finding Gravity within Quantum Mechanics

For more on the implications and limitations of locality, see G. Musser (2015), *Spooky Action at a Distance: The Phenomenon That Reimagines Space and Time—and What It Means for Black Holes, the Big Bang, and Theories of Everything*, Farrar, Straus and Giroux.

"I use more brain grease": A. Einstein, quoted by Otto Stern (1962), interview with T. S. Kuhn, Niels Bohr Library & Archives, American Institute of Physics, https://www.aip.org/history-programs/niels-bohr-library/oral-histories/4904.

"Perhaps the success of the Heisenberg method": A. Einstein (1936), "Physics and Reality," reprinted in A. Einstein (1956), *Out of My Later Years*, Citadel Press.

T. Jacobson (1995), "Thermodynamics of Space-Time: The Einstein Equation of State," *Physical Review Letters* 75, 1260.

T. Padmanabhan (2010), "Thermodynamical Aspects of Gravity: New Insights," *Reports on Progress in Physics* 73, 046901.

E. P. Verlinde (2011), "On the Origin of Gravity and the Laws of Newton," *Journal of High Energy Physics* 1104, 29.

J. S. Cotler, G. R. Penington, and D. H. Ranard (2019), "Locality from the Spectrum," *Communications in Mathematical Physics*, https://doi.org/10.1007/s00220-019-03376-w.

J. Maldacena and L. Susskind (2013), "Cool Horizons for Entangled Black Holes," *Fortschritte der Physik* 61, 781.

C. Cao, S. M. Carroll, and S. Michalakis (2017), "Space from Hilbert Space: Recovering Geometry from Bulk Entanglement," *Physical Review D* 95, 024031.

C. Cao and S. M. Carroll (2018), "Bulk Entanglement Gravity Without a Boundary: Towards Finding Einstein's Equation in Hilbert Space," *Physical Review D* 97, 086003.

T. Banks and W. Fischler (2001), "An Holographic Cosmology," https://arxiv.org/abs/hep-th/0111142.

S. B. Giddings (2018), "Quantum-First Gravity," *Foundations of Physics* 49, 177.

D. N. Page and W. K. Wootters (1983). "Evolution Without Evolution: Dynamics Described by Stationary Observables," *Physical Review D* 27, 2885.

Chapter 14

Beyond Space and Time

Holography, Black Holes, and the Limits of Locality

Holography, complementarity, and black hole information are discussed in L. Susskind (2008), *The Black Hole War: My Battle with Stephen Hawking to Make the World Safe for Quantum Mechanics*, Back Bay Books.

A. Almheiri, D. Marolf, J. Polchinski, and J. Sully (2013), "Black Holes: Complementarity or Firewalls?" *Journal of High Energy Physics* 1302, 062.

J. Maldacena (1997), "The Large-N Limit of Superconformal Field Theories and Supergravity," *International Journal of Theoretical Physics* 38, 1113.

S. Ryu and T. Takayanagi (2006), "Holographic Derivation of Entanglement Entropy from AdS/CFT," *Physical Review Letters* 96, 181602.

B. Swingle (2009), "Entanglement Renormalization and Holography," *Physical Review D* 86, 065007.

M. Van Raamsdonk (2010), "Building Up Spacetime with Quantum Entanglement," *General Relativity and Gravitation* 42, 2323.

Epilogue

Everything Is Quantum

"A wonder of this kind": A. Einstein (1949), *Autobiographical Notes*, Open Court Publishing, 9.

Appendix

The Story of Virtual Particles

For more on Feynman diagrams, see R. P. Feynman (1985), *QED: The Strange Theory of Light and Matter*, Princeton University Press.

INDEX

ABOUT THE AUTHOR

SEAN CARROLL is a theoretical physicist at the California Institute of Technology. His research has focused on cosmology, gravitation, field theory, quantum mechanics, statistical mechanics, and foundations of physics. He has received numerous awards, including the Andrew Gemant Award from the American Institute of Physics, the Royal Society Prize for Science Books, and a Guggenheim Fellowship. His other books include *From Eternity to Here: The Quest for the Ultimate Theory of Time*; *The Particle at the End of the Universe: How the Hunt for the Higgs Boson Leads Us to the Edge of a New World*; and *The Big Picture: On the Origins of Life, Meaning, and the Universe Itself*. He also hosts the weekly *Mindscape* podcast.

www.preposterousuniverse.com